PRINCIPLES OF ANIMAL AGRICULTURE

PRINCIPLES OF ANIMAL AGRICULTURE

Charles E. Stufflebeam

Professor of Animal Science
Southwest Missouri State University

PRENTICE-HALL, INC., ENGLEWOOD CLIFFS, N.J. 07632

Library of Congress Cataloging in Publication Data

Stufflebeam, C. E. (Charles E.)
  Principles of animal agriculture.

  Bibliography: p.
  Includes index.
    1. Livestock. I. Title.
SF61.S78   636   82–3838
ISBN 0-13-700948-8   AACR2

PRINCIPLES OF ANIMAL AGRICULTURE
Charles E. Stufflebeam

Editorial/Production Supervision: Karen J. Clemments
Interior Design: Karen J. Clemments
Chapter Opening Design: Janet Schmid
Cover Design: Janet Schmid
Manufacturing Buyer: John Hall

Printed in the United States of America

10 9 8 7 6 5 4

ISBN 0-13-700948-8

Prentice-Hall International, Inc., *London*

Prentice-Hall of Australia Pty. Limited, *Sydney*

Prentice-Hall of Canada Inc., *Toronto*

Prentice-Hall of India Private Limited, *New Delhi*

Prentice-Hall of Japan, Inc., *Tokyo*

Prentice-Hall of Southeast Asia Pte. Ltd., *Singapore*

Whitehall Books Limited, *Wellington, New Zealand*

Contents

Preface

This book presents principles of animal agriculture essential for a basic understanding of the animals that are the chief producers of food and fiber for human consumption. When I wrote this book, I tailored the content to meet the needs of the beginning student and of the instructor as well.

Almost one-half the men and women studying agriculture today have never lived on a farm nor have they studied agriculture at the secondary school level. Even those who have some agricultural experience are not familiar with all aspects of the agricultural industries or with the biological principles associated with domestic animals. The chapters on specific breeds and animal behavior should be of particular interest to these readers. The biological principles of anatomy, physiology, genetics, reproduction, nutrition, and health are presented throughout the book and require no prior preparation in basic biology beyond the most elementary knowledge.

The concise, uncluttered nature of this book lends itself to use in courses at many levels. A course on the fundamentals of animal agriculture, such as those in vocational agricultural programs, can be taught on a chapter-by-chapter basis. A more intensive course, typically found in a community college or university, can be taught by referring the student to sources listed in the bibliography at the end of the book. Entries are listed according to topic and they include the well-known titles in each field.

In order to assist students in their understanding and retention of important principles, the book is arranged as follows: Each chapter begins with a study outline that follows the organization of the chapter material. This outline is helpful as a preview to reading as well as a guide to review later. Illustrations and tables clarify the topics discussed and provide additional information. Illustrations are kept simple and straightforward intentionally in order to enhance understanding.

A knowledge of terminology is basic to a complete understanding of each area of animal agriculture. For this reason, every key term is described and defined the first time it is used. These terms are listed again at the end of the chapter in the form of a question. The complete glossary at the end of the book also includes the key terms and their definitions.

An understanding of the concepts and information presented in this book will lead to a better and fuller appreciation of animal agriculture as a segment of the whole agricultural industry. This book will help the reader become aware of the tremendous impact that agriculture has on the lives of people all over the world. It will also provide the basic language needed to communicate with other agriculturists, but, more particularly, to communicate with those outside the agricultural community.

For students who will choose a career in animal agriculture, this book will provide the foundation for advanced studies in nutrition, genetics, reproduction, diseases, physiology, products, and other segments of animal agriculture.

C. E. S.

PRINCIPLES OF ANIMAL AGRICULTURE

1

The Scope of Animal Agriculture

Agriculture involves the growing of plants and animals for food, fiber, and other human needs. It could well be classified as the most important industry in the world. The meaning of the word agriculture has changed somewhat since it was first used. It comes from two Latin words that mean to "cultivate the field." Plants were probably cultivated for food long before animals were domesticated, so the first agriculture was essentially limited to the growing of food crops.

As animals were domesticated and grown for food purposes, the meaning of agriculture broadened to include animal husbandry. Husbandry involves the care and management of plants or animals. The root of the word is the same as that for husband, a man who takes care of his household. Today, the subject of agriculture includes the various industries and services that are directly involved with agricultural products, and those that depend upon agriculture for their existence. Among these related areas are farm machinery, chemicals, education, meat packing, feeds, marketing, communication services, and many others.

DEVELOPMENT OF ANIMAL AGRICULTURE

Language of Animal Agriculture

The study of animals in agriculture can be referred to by any one of several terms. One term we have already used is animal husbandry. Traditionally, animal husbandry referred only to the meat-producing animals and to horses, and did not include poultry and dairy cattle. In recent years, however, the term animal husbandry as a category of agriculture has been giving way to terms such as animal science and animal industry, which generally include all farm animals. Farm animals are often referred to as livestock. Although some people use this word to refer only to the meat-producing animals, in this book the term livestock refers to all species of farm animals.

Animal agriculture by whichever term you choose—animal science, animal industry, or animal husbandry—can be subdivided into several areas for additional study. One method of subdivision is based on the various biological functions of the animal. It is this approach that has been used to a large extent in the organization of this book. Those body functions having a more direct relationship to the animal's economic importance will receive greater emphasis. These important areas include genetics, reproduction, nutrition, disease resistance, and production of food products.

Animal science can also be studied according to species and classes. Traditionally, this has been the way that much of the study of animals has been carried out. We thus have the areas of poultry science or poultry husbandry, sheep husbandry, swine science, horse science, dairy cattle husbandry, and beef cattle science. Further specialization can lead to the study of dairy cattle nutrition, swine marketing, horse breeding, poultry genetics, dairy products, meat science, and so on.

As the study of animal agriculture progresses, more and more new terms will be encountered. This terminology must be learned in order to understand and be able to communicate the subject matter involved. As each new subject is presented in the succeeding chapters,

TABLE 1-1.   TERMINOLOGY FOR THE VARIOUS
                  AGE AND SEX GROUPS OF DOMESTIC
                  ANIMALS

Species	Mature Male	Mature Female	Newborn of Either Sex	Young Male	Young Female	Castrate Male
Cattle	Bull	Cow	Calf	Bull	Heifer	Steer
Horse	Stallion	Mare	Foal	Colt	Filly	Gelding
Swine	Boar	Sow	Pig	Boar	Gilt	Barrow
Sheep	Ram	Ewe	Lamb	Ram	Ewe	Wether
Goat	Buck	Doe	Kid	Buck	Doe	Wether
Chicken	Cock	Hen	Chick	Cockerel	Pullet	Capon
Duck	Drake	Duck or hen	Duckling	—	—	—
Goose	Gander	Goose or hen	Gosling	—	—	—
Turkey	Tom	Hen	Poult	—	—	—

many new terms will be presented and defined. One of the purposes of this first chapter is to introduce some of the basic terminology dealing with animals. Among the most basic terms are those given in Table 1-1, which lists several species of domestic animals and the proper terminology for the various age and sex groups.

In addition to knowing the proper terminology for the sex and age groups of the farm animals, the student of animal agriculture should also be familiar with the names of the various external parts of these animals. Figures 1-1, 1-2, and 1-3 show diagrams of the horse, pig, cow, sheep, turkey, and chicken with some of the body parts labeled.

Domestication of Animals

As far as contemporary production of food and fiber is concerned, the history of the domestication of animals is probably not of extreme importance. Although we know that domestication of our common farm animals occurred before history was recorded, we cannot be certain how and when it came about. Most of the evidence that is available comes from archeological studies.

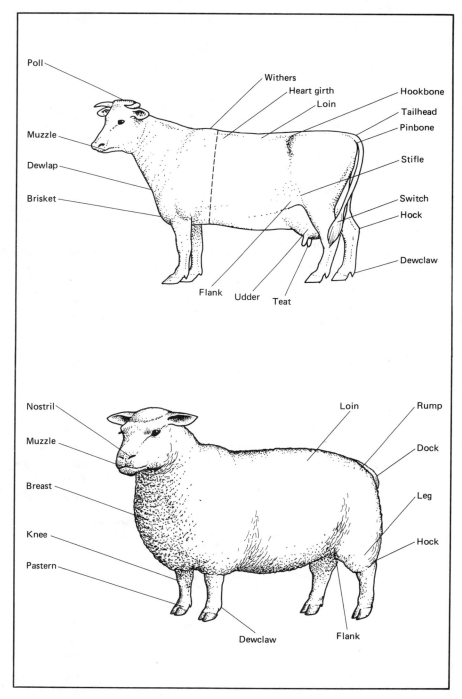

Figure 1-1   External parts of a cow and a sheep.

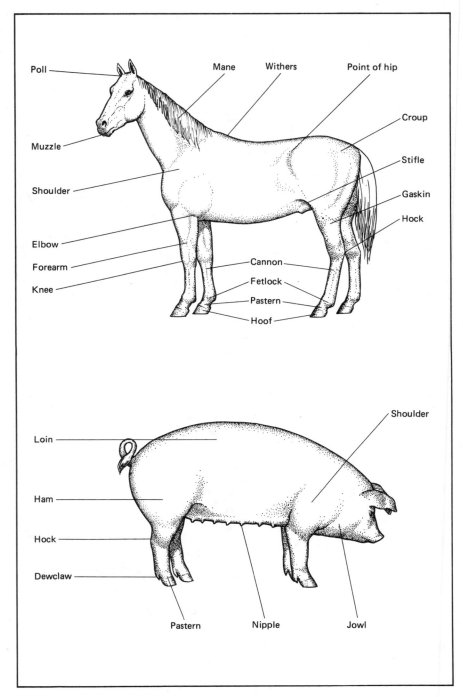

Figure 1-2   External parts of a horse and a pig.

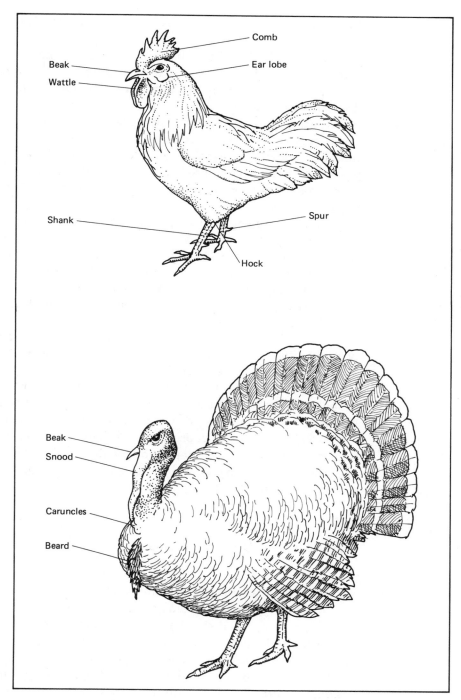

Figure 1-3 External parts of a chicken and a turkey.

The dog is thought to have been the first animal to be domesticated. This first instance of domestication may have occurred as long as 12,000 years ago in the region that is now Iraq. The dog was domesticated from the wolf, probably for reasons of companionship, but possibly also for food. By this time in the life of the human race, certain crops were probably already under cultivation. It is thought that our ancestors began some cultivation of plants about 15,000 years ago. Thus, compared to the several million years that the human race has been in existence, cultivation and domestication are relatively recent events.

Somewhere within the time period of 9000 to 11,000 years ago, sheep, goats, swine, and cattle were domesticated. Sheep were the first of these four species to be domesticated, then goats and swine, and then cattle. Swine were domesticated independently but at about the same time in China, Egypt, and parts of Europe. The other three species were probably domesticated in the region around the northeastern Mediterranean.

Domestic chickens came on the scene in the region of India and possibly other areas some 5000 to 6000 years ago. As far back as 4000 years ago, horses were used to pull chariots in time of war. Donkeys were used to carry burdens and their domestication is thought to have occurred in Egypt. Cats and guinea fowl were probably also domesticated in Africa. The time of domestication for the turkey, duck, and goose is not known for sure. Turkeys were domesticated in North America, ducks in China, and geese in Greece and Italy. A summary of the times and places of animal domestication is presented in Table 1-2.

Classification of Domestic Animals

The scientific classification system for animals can be found in almost any zoology or general biology text. All living things are classified into large groups called kingdoms. For many years there were only the plant and animal kingdoms. In recent years, biological taxonomists have proposed the classification of living things into at least three and as many as five kingdoms. In the newer systems of classification, bacteria and fungi are not included in the plant kingdom as

TABLE 1-2.   APPROXIMATE TIMES AND REGIONS
OF DOMESTICATION OF ANIMALS

Species	Approximate Number of Years Since Domestication	Region
Dog	12,000	Iraq
Sheep	11,000	Old World
Goat	11,000	Old World
Pig	10,000	China, Egypt, Europe
Cattle	9000	Old World
Chicken	6000	India
Horse	5000	Old World
Donkey	5000	Egypt
Cat	4000	Egypt
Turkey	?	North America

they formerly were, and protozoa are excluded from the animal kingdom.

A kingdom is subdivided into large groups called phyla. A phylum is subdivided into classes, a class into orders, an order into families, a family into genera, and a genus into species. If any of these major categories need to be subdivided further, we find subphyla, suborders, and subfamilies being used. Depending upon the particular species being classified, species may be further subdivided into varieties, breeds, strains, or lines.

All domestic animals are members of the kingdom Animalia, phylum Chordata, subphylum Vertebrata, and either the class Mammalia or the class Aves. Mammals grow hair and produce milk for their young. Birds (avians) grow feathers, lay eggs, have no teeth, and do not have a diaphragm. The diaphragm is a thin muscular tissue that separates the thoracic and abdominal cavities in mammals. Both birds and mammals are homeothermic; that is, they have the ability to maintain a constant body temperature. Thus, they are generally referred to as warm-blooded animals. All other animals, such as reptiles, fishes, insects, and worms, are poikilothermic, or cold-blooded. Their body temperatures vary with that of the environment in which they live.

Figure 1-4 lists the classification scheme for selected domestic animals. The rules of taxonomy (the science of classifying living things) prescribe that the scientific name of an animal include both the genus and the species. The term for the species is not capitalized, but all other scientific terms are. Further, the names of the genus and species are either underscored or printed in italicized letters.

A species is defined as a group of animals with certain common characteristics that, when mated among themselves will produce

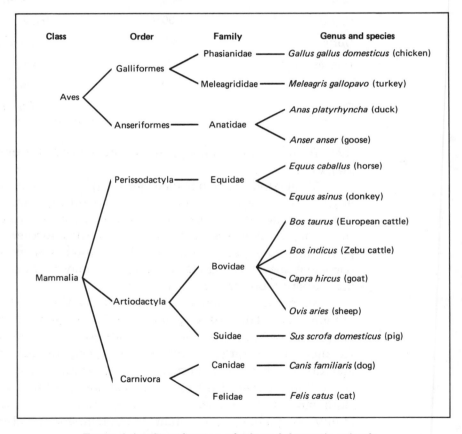

Figure 1-4   Classification of selected domestic animals.

fertile offspring. By this definition *B. taurus* (European cattle) and *B. indicus* (Indian cattle) are technically members of the same species. The American bison are members of the family Bovidae and also members of the genus *Bos* by some classification schemes. Their scientific name is *B. bison*. Bison and cattle will sometimes interbreed and produce fertile offspring. The mule is a member of the genus *Equus*, but is a hybrid of the species *E. caballus* and *E. asinus*, actually a cross between a donkey jack and a horse mare. A donkey jenny mated to a stallion would produce a hinny.

Taxonomists do not always agree as to how some animals should be classified. For example, in some references the domestic pig is referred to by the scientific name *S. scrofa*. Other references will use the term *S. domesticus*, and one reference uses the term *S. scrofa domesticus*. The wild pig from which our pigs were domesticated is called *S. scrofa*. The process of domestication does not create a new species, so the latter terminology may be quite appropriate.

ANIMAL AGRICULTURE AROUND THE WORLD

Human Population Trends

During the time of Christ there were about 250 million people in the world. By the year 1600 the population had doubled to approximately one-half billion. Between 1600 and 1830, the population increased by another one-half billion. In other words, in only 230 years the population had doubled once again, reaching a total of 1 billion persons. The next doubling of the population required only 100 years—by 1930 there were 2 billion people on the planet. By 1960, just 30 years later, another 1 billion people had been added, bringing the total to 3 billion. Between 1930 and 1975, a period of 45 years, the population had doubled from 2 to 4 billion.

Various projection estimates by the Food and Agriculture Organization of the United Nations (FAO) and others suggest that there will be more than 6 billion people in the world by the year 2000. If this estimate turns out to be correct, the population will have doubled in only 40 years from 3 billion in 1960. Figure 1-5 shows in graphic form how the population of the world has been increasing since the time of Christ. Can you understand why some people refer to this tremendous change as a population explosion?

The countries of the world have been divided into two groups based upon their degree of industrial development and ability to

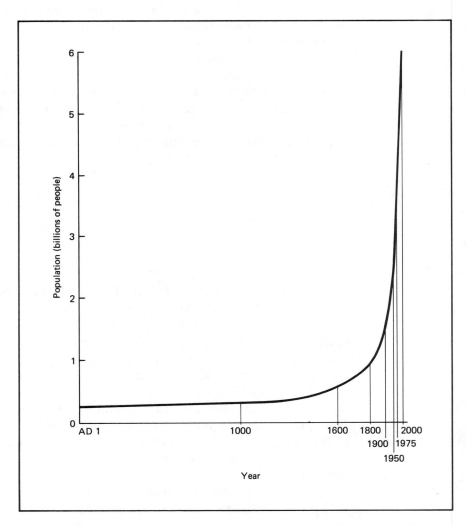

Figure 1-5   Trend in the growth of the human population since the time of Christ.

provide food for their citizens. The developed countries include those in North America, Europe, the Soviet Union, Japan, and a few others. The less-developed, or developing, countries include those in Africa, much of Asia, and Latin America. Less than 30 percent of the population of the world is found in the developed countries. Population increases in these countries are taking place at the rate of about 1 percent per year. However, the population in the developing countries is increasing at the rate of almost 2.5 percent per year. At these rates, the population of the developed countries will be down to 25 percent of the total world population by the year 2000.

World Nutrition Problems

In the developing countries, where 70 percent of the people of the world live, only about 40 percent of the world's food supply is being produced. In many of these countries, the amount of protein and energy consumed by the people is below normal. In addition, in many of these same countries the proportion of the total protein consumed that comes from animal sources is very low. This fact is significant because animal protein is generally much better in providing human nutritional needs than is plant protein. Proteins found in eggs, milk, meat, and fish are referred to as complete proteins because they contain all the amino acid building blocks that are essential for replacing and building body protein. Plant proteins, on the other hand, are generally deficient in one or more of these important building blocks. Soybean protein is an exception; it is a complete protein. A more complete explanation of proteins and amino acids will be given in Chapter 11. Meanwhile, it is important to note that approximately one-third of a person's total protein requirement should come from animal sources.

The low level of food production in the developing countries, along with their higher rate of population increase, is compounding the problem of providing for the nutritional needs of the people who live in these countries. The problems of reducing birth rates and increasing the supplies of more nutritious food materials are not simple ones. It would seem that with our knowledge of technology and economics, and with the help of education and transportation, such problems could be solved. Yet, because of circumstances that relate to politics, religion, and culture in some of the developing countries, it may well be that the problems of population and hunger will never be completely solved. However, even though these problems may seem unsolvable, this does not mean that we should not make an attempt to alleviate the conditions that underlie them.

Production of Livestock and Their Products in Selected Countries

The data in Tables 1-3 and 1-4 are presented to show where agricultural animals and their products are found in the world. Although the specific numbers change from year to year, they usually do not change to a very great extent. Countries with the greatest populations and the largest geographical areas would be expected to rank high in numbers of certain kinds of animals and their products, and generally this is true.

About one-fourth of the total cattle in the world are found in India. India, however, produces very little beef and milk. Because of religious beliefs in that country, very few cattle are slaughtered there. Most of the cattle produced for meat purposes are found in the United States, the Soviet Union, and the Latin American countries. France ranks among the top ten countries in total number of cattle, but well over 40 percent of the cattle of that country are used for milk purposes. By comparison, about 10 percent of the cattle in the United States are dairy cattle and about 5 percent in the world as a whole. In fact, France ranks second only to the United States in numbers of dairy cattle and in milk production. Dairy cattle and milk production are concentrated in the European countries and in the United States and Canada.

The greatest numbers of swine, or hogs, are found in China, the Soviet Union, the United States, and Brazil. Hog production is quite concentrated in the countries of Central and Eastern Europe, also. The United States ranks first in the total tonnage of pork produced because of a greater production rate per head. However, West Germany and France each produce more pork per head than the United States, suggesting that in those countries, hogs are carried to heavier slaughter weights than in the United States.

The two leading countries in total sheep numbers are also the two leaders in wool, lamb, and mutton production. However, Australia ranks first in wool production while ranking behind the Soviet

TABLE 1-3. PRODUCTION OF LIVESTOCK AND WOOL IN SELECTED COUNTRIES

| Total Cattle (1980) | | Milk Cows (1979) | | Swine (1980) | |
Country	Thousands	Country	Thousands	Country	Thousands
1. India	241,000	1. United States	10,777	1. China	250,000[a]
2. USSR	115,000	2. France	10,222	2. USSR	73,700
3. United States	110,961	3. West Germany	5360	3. United States	66,950
4. Brazil	93,000	4. United Kingdom	3433	4. Brazil	36,800
5. Argentina	59,245	5. Italy	3045	5. West Germany	22,361
6. Mexico	29,500	6. Netherland	2325	6. Poland	21,850
7. Columbia	27,151	7. New Zealand	2040	7. Mexico	12,850
8. Australia	26,100	8. Australia	1970	8. East Germany	11,775
9. France	23,834	9. Canada	1870	9. Romania	11,700
10. Turkey	16,200	10. Ireland	1545	10. Spain	11,500
World	948,377	World	49,075	World	681,535

TABLE 1-3. *(Continued)*

| Chickens (1979) | | Sheep (1980) | | Wool (1980) | |
Country	Millions	Country	Thousands	Country	Thousands of Metric Tons
1. China	1250[a]	1. USSR	143,650	1. Australia	710
2. USSR	430	2. Australia	135,000	2. USSR	460
3. United States	291	3. China	75,000[a]	3. New Zealand	320
4. Japan	123	4. New Zealand	65,000	4. Argentina	170
5. Poland	74	5. Turkey	44,500	5. South Africa	111
6. France	70	6. India	40,500	6. Uruguay	63
7. West Germany	58	7. Argentina	35,300	7. China	61
8. United Kingdom	57	8. Iran	35,000[a]	8. Turkey	55
9. Turkey	51	9. South Africa	32,320	9. United Kingdom	54
10. Italy	50	10. United Kingdom	23,500	10. United States	46
World	2897	World	780,143	World	2529

Source: *Agricultural Statistics*, 1980. USDA.
[a]Estimated.

TABLE 1-4. PRODUCTION OF MEAT, MILK, AND EGGS IN SELECTED COUNTRIES, 1979

Beef		Pork		Lamb, Mutton, and Goat Meat	
Country	Thousands of Metric Tons	Country	Thousands of Metric Tons	Country	Thousands of Metric Tons
1. United States	9925	1. United States	7008	1. USSR	855
2. USSR	6510	2. USSR	3656	2. Australia	528
3. Argentina	3085	3. West Germany	2690	3. New Zealand	505
4. Brazil	2100	4. Poland	1842	4. Iran	377[a]
5. France	1832	5. France	1690	5. India	370
6. Australia	1808	6. Japan	1430	6. Turkey	318
7. West Germany	1520	7. Netherlands	1040	7. United Kingdom	225
8. Italy	1075	8. East Germany	990	8. South Africa	174
9. Mexico	1037	9. Italy	921	9. France	168
10. Canada	955	10. United Kingdom	910	10. Spain	125
World	39,643	World	32,234	World	4320

TABLE 1-4. *(Continued)*

	Total Meat		Milk		Eggs	
	Country	Thousands of Metric Tons	Country	Thousands of Metric Tons	Country	Millions
1.	United States	17,066	1. United States	56,074	1. United States	69,227
2.	USSR	11,021	2. France	31,560	2. USSR	65,000
3.	West Germany	4245	3. West Germany	23,800	3. Japan	33,500
4.	France	3730	4. United Kingdom	15,340	4. West Germany	14,000
5.	Argentina	3504	5. Netherlands	11,580	5. United Kingdom	13,950
6.	Brazil	3093	6. Italy	11,100	6. France	13,800
7.	Poland	2618	7. Canada	7675	7. Spain	11,194
8.	Australia	2541	8. Japan	6490	8. Italy	11,100
9.	United Kingdom	2178	9. New Zealand	6361	9. Mexico	10,950
10.	Italy	2102	10. Australia	5400	10. Poland	8980
	World	76,514	World	198,015	World	321,877

Source: *Agricultural Statistics,* 1980. USDA
[a] 1978 data.

Union in the production of total sheep and meat. Outside the Oceanic countries of Australia and New Zealand, sheep seem to be heavily concentrated in the region of the Persian Gulf. In Iran, Turkey, and India, sheep seem to be raised primarily for meat purposes rather than wool. In these nations, wool production rate per animal is very low compared to that in other countries. The ratio of wool production to number of sheep is highest in Australia, Argentina, and New Zealand, followed by the United States. The ratio of lamb and mutton production to number of sheep is highest in France and the United States, followed by the United Kingdom, Spain, and the Persian Gulf countries. The United States ranks among the top ten countries in both the production of wool and meat from sheep, but ranks about seventeenth in total number of sheep.

The United States ranks first in the production of eggs even though it is only third in terms of the number of chickens produced. This reflects a greater egg production rate per bird compared to other countries. Even though Japan is a rather small country in geographical area, it is one of the leaders in the production of poultry. Because of the greater concentration of people in that country, it is much easier to raise chickens than cattle.

ANIMAL AGRICULTURE IN THE UNITED STATES

Trends in Livestock Production

A study of Figure 1-6 reveals some interesting changes that have occurred in numbers of livestock in the United States during the past 60 years.

The overall trend in the total number of cattle has been upward, but the growth curve has some peaks and valleys. Since 1934, there has been a peak in cattle numbers followed by a decline about every 10 years. These peaks occurred in 1945, 1955, 1965, and 1975. The peak in the cycle in 1965 was a little different in that, instead of declining, cattle numbers remained about the same for 2 or 3 years and then started climbing again. This tendency for cattle numbers to increase and then decline is referred to as the cattle cycle. Before 1934, the average length of the cattle cycle was about 14 years.

Most of the effect upon the overall cattle cycle has been caused

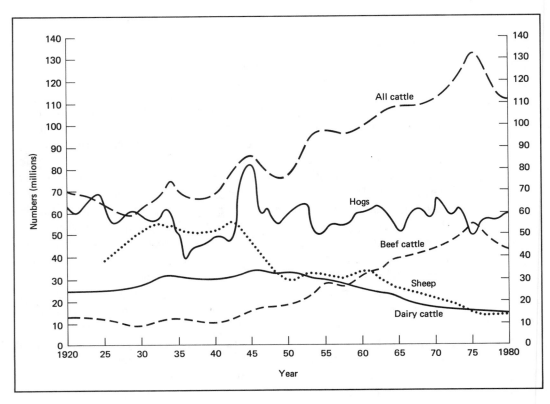

Figure 1-6 Livestock numbers in the United States since 1920.

by fluctuations in the numbers of beef cattle rather than dairy cattle. Because of the greater investment in facilities and equipment in the dairy business, it is not an easy matter to increase and decrease the size of the dairy herd when milk prices change. From about 1945 onward there has been a steady annual decline in dairy cattle numbers. During this period, the number of dairy cows has been reduced by more than one-half, while the total milk production has only decreased by about 7 or 8 percent. The production rate per cow during this period has more than doubled.

The decrease in sheep numbers during the past 35 years has been about as dramatic as has been the increase in beef cattle numbers. During this period, beef cow numbers have increased about five-fold while sheep numbers have been reduced to one-fifth of their former level. Sheep numbers peaked in 1942 at about 55 million head. By 1950 their numbers had dropped to below 30 million. Sheep numbers remained rather steady during the 1950s, and even increased to about 33 million head by 1960. Since 1960, the decline in numbers has been quite steady until now there are about 10 million sheep in the country.

A number of factors seem to have been involved in the decline of sheep numbers. One of these factors has been the development of more economical ways of producing textiles synthetically. This has resulted in a decrease in the demand for wool. Compared to hogs and cattle, there are more complicated management problems associated with the raising of sheep. They are more susceptible to certain diseases and predators, and require more attention at lambing time. The demand for the meat of sheep—lamb and mutton—has decreased rather significantly, also.

There have been many fluctuations in swine numbers during the past 60 years, but the overall trend in numbers has been neither up nor down. The population curve for swine has been fluctuating above and below the 60 million mark since 1920. The total number of swine reached a low of around 40 million during the Great Depression of the 1930s. Within 10 years, however, at the end of World War II, this number had more than doubled to an all-time high of over 80 million.

From 1945 until very recently, there have been ten peaks in swine numbers. This suggests that the average length of the swine cycle is about three and one-half years. Because of their great reproductive abilities and their short generation interval, it is possible to increase swine numbers very rapidly. A rapid increase in the number of hogs soon results in a lower price at the markets, followed by a decrease in hog numbers again. Hog raisers will experience about three such cycles to every one for the person raising beef cattle.

As in the case of milk production, the number of eggs produced per hen per year has increased considerably during the past 50 years. In 1935, the annual production per hen was about 120 eggs. That statistic is now up to about 230 eggs per year. This improvement has come about through several factors including genetic selection and advances in nutrition and disease control. The use of artificial lighting to adjust the length of the day to correspond more closely to the length of the egg-laying cycle has also been an important factor. The number of laying hens and total eggs produced has declined sharply in recent years. Since eggs are a major food source of cholesterol, this decline may be tied in with medical concerns over cholesterol and its possible connection with heart disease.

The numbers of both broilers and turkeys have increased tremen-

dously during the past 40 years or so. Broiler production in the
United States has increased about 100-fold since 1930. Turkey production has increased about 50-fold. Today most of the broilers and
turkeys are produced under contracts with large processing firms or
feed companies. This kind of system where processors and feed companies own the birds they process or feed is an example of vertical
integration.

Areas of Livestock Production

It is probably no surprise to anyone that the state of Texas leads all
other states in the numbers of beef cattle, as well as in total cattle.
The ten leading states for each of these groups of cattle along with
the leading states in dairy cattle numbers are presented in Table 1-5.

The list of the leading states in beef cow numbers indicates that
cow-calf operations are most numerous in the midwestern states of
Missouri, Oklahoma, Nebraska, Iowa, and Kansas. If cow numbers
were expressed on a per-square-mile basis, Texas would still be in
the top ten, but would not be number one. This concentration of
beef cattle production extends into the southern and southeastern
part of the country as well as into Texas. The number of beef cows
in the ten leading states accounts for about 55 percent of the total
for the country.

Wisconsin is the leading state in total number of dairy cattle.
This one state accounts for about one-sixth of all the dairy cattle
in the country. The six states along the northern border of the
United States from Wisconsin to New York have almost one-half of
all the dairy cattle in the country. The area encompassing these six
states plus Iowa and Missouri could very well be called the Dairy Belt
of the United States. Because of the large number of dairy cattle in
Wisconsin, it also ranks among the top ten states in total number of
cattle.

One-fourth of all the hogs in the United States are in the state
of Iowa. One-half of the total number of hogs in the country are
found in the four leading states. The Swine Belt of the country
coincides very closely with the Corn Belt. About three-fourths of the
nation's corn crop is usually fed to livestock of which about one-third
is fed to hogs. The leading states in numbers of hogs, sheep, and
poultry are presented in Table 1-6.

TABLE 1-5. LEADING STATES IN CATTLE NUMBERS, JANUARY 1980

Beef Cows and Heifers		Dairy Cows and Heifers		Total Cattle	
State	Thousands	State	Thousands	State	Thousands
1. Texas	6495	1. Wisconsin	2548	1. Texas	13,200
2. Missouri	2547	2. New York	1268	2. Iowa	7150
3. Oklahoma	2460	3. Minnesota	1250	3. Nebraska	6400
4. Nebraska	2238	4. California	1215	4. Kansas	6100
5. Kansas	1946	5. Pennsylvania	978	5. Oklahoma	5500
6. Iowa	1932	6. Michigan	593	6. Missouri	5350
7. South Dakota	1728	7. Ohio	531	7. California	4550
8. Montana	1702	8. Iowa	515	8. Wisconsin	4280
9. Florida	1280	9. Texas	405	9. South Dakota	4010
10. Kentucky	1268	10. Missouri	372	10. Minnesota	3750
Ten State Total	23,596	Ten State Total	9675	Ten State Total	60,290
United States	42,914	United States	14,976	United States	110,961

Source: *Agricultural Statistics*, 1980. USDA.

TABLE 1-6. NUMBERS OF SWINE, SHEEP AND POULTRY IN SELECTED STATES

Swine		Sheep		Turkeys	
State	Thousands[a]	State	Thousands[b]	State	Thousands[c]
1. Iowa	16,200	1. Texas	2400	1. Minnesota	24,666
2. Illinois	6950	2. California	1175	2. North Carolina	23,100
3. Indiana	4900	3. Wyoming	1050	3. California	18,855
4. Minnesota	4900	4. Colorado	870	4. Arkansas	13,340
5. Missouri	4550	5. South Dakota	783	5. Missouri	10,950
6. Nebraska	4150	6. New Mexico	660	6. Virginia	9174
7. North Carolina	2600	7. Utah	507	7. Texas	8000
8. Georgia	2080	8. Montana	494	8. Iowa	6160
9. Ohio	2070	9. Idaho	468	9. Wisconsin	5645
10. Kansas	2070	10. Arizona	400	10. Indiana	5640
United States	66,950	United States	12,513	United States	156,529

TABLE 1-6. *(Continued)*

Chickens		Laying Hens		Broiler Chickens	
State	Thousands[a]	State	Thousands[a]	State	Thousands[a]
1. California	45,510	1. California	17,794	1. Arkansas	678,208
2. Georgia	36,200	2. Florida	9921	2. Georgia	561,268
3. Arkansas	27,000	3. Georgia	9100	3. Alabama	493,060
4. Alabama	22,000	4. Arkansas	6650	4. North Carolina	376,580
5. Pennsylvania	20,930	5. Indiana	6150	5. Mississippi	276,858
6. North Carolina	20,800	6. Pennsylvania	5230	6. Maryland	244,783
7. Indiana	20,500	7. Alabama	5000	7. Texas	217,461
8. Florida	18,530	8. Texas	4300	8. Delaware	175,856
9. Texas	17,500	9. Minnesota	3813	9. California	137,600
10. Minnesota	12,500	10. North Carolina	3800	10. Virginia	111,564
United States	399,676	United States	119,997	United States	3,939,832

Source: *Agricultural Statistics*, 1980. USDA.

[a] As of December 1979.
[b] As of January 1980.
[c] As of August 1979.

About one-fifth of all the sheep in the United States are found in Texas. The majority of these animals are concentrated in the western part of the state. Except for California, the ten leading states in sheep numbers are western range states. A typical sheep operation in the range states will have over 1000 ewes running on vast areas of unfenced land. In addition, scattered throughout the country, are farm flocks ranging in size from just a few ewes to several hundred.

California is the leading state in total number of chickens as well as the leading egg-producing state. The leading broiler-producing state is Arkansas. With these two exceptions, the broilers and laying hens are quite heavily concentrated in the southeastern part of the United States. There does not appear to be any clearly defined area of turkey production. The leading turkey-producing states are situated on the East Coast, on the West Coast, and in the north central part of the country.

STUDY QUESTIONS AND EXERCISES

1. Define each of the following terms:

   agriculture

   avian

   Bovidae

   buck

   diaphragm

   domesticate

   doe

   drake

   duckling

   gander

   homeothermic

   husbandry

   gosling

   hinny

   jack

   jenny

   livestock

   mammal

   mule

   poikilothermic

   species

   vertical integration

   taxonomy

2. Explain why agriculture could be considered as the world's leading industry.

3. How has the meaning of agriculture changed over the years?

4. Name some of the terms by which the study of animal agriculture is known.

5. Identify and locate the various parts of the farm animals illustrated in Figures 1-1, 1-2, and 1-3.

6. Which of the domestic animals was probably the first to be domesticated and for what purpose? Which was the last?

7. During what period of time were horses, cattle, sheep, hogs, and chickens each domesticated, and where?

8. Which occurred first, domestication of animals or cultivation of crops?

9. Identify the age group, sex group, and species of each of these terms: barrow, colt, doe, filly, gilt, and poult.

10. Distinguish between mammals and birds.

11. Identify each of the farm animals as to order, family, genus, and species.

12. In terms of scientific classification, describe the mule.

13. Discuss whether or not it is appropriate to classify European cattle and Indian cattle as separate species.

14. Describe the trend in population growth of humans since the time of Christ.

15. What is meant by the term population explosion?

16. Give some examples of countries that are considered to be developed countries.

17. What are some examples of developing countries?

18. Compare the human population and agricultural production in the developed and developing countries.

19. Why are the proteins from eggs, milk, meat, fish, and soybeans called complete proteins?

20. Why are many of the people in the developing countries undernourished?

21. What are some factors that could help reduce the hunger problems in the developing countries?

22. List the three leading countries in numbers of dairy cattle, total cattle, hogs, sheep, and chickens.

23. How does the United States rank with the other countries of the world in numbers of dairy cattle, total cattle, hogs, sheep, and chickens?

24. List the three leading countries in total production of beef, pork, lamb, total meat, milk, eggs, and wool.

25. How does the United States rank with the other countries in the production of beef, pork, lamb, total meat, eggs, and wool?

26. Describe the trends in numbers for each class of farm animal in the United States over the past 50 years.

27. Describe what is meant by the cattle cycle.

28. Compare the cattle cycle with the swine cycle.

29. Why are swine numbers able to fluctuate so much from year to year compared to dairy cattle numbers?

30. Explain how dairy cattle numbers can be reduced by 50 percent over a certain period of time while total milk production may only decrease by 10 percent.

31. In what general area of the country are most of the hogs found?

32. What has happened to numbers of chickens and turkeys compared to their production rates during the past 50 years?

33. Generally, in what part of the United States are dairy cattle found?

34. Name the three leading states in numbers of beef cows, dairy cows, total cattle, hogs, sheep, turkeys, laying hens, and broilers.

2

Breeds
of Farm Animals

A breed is a group of animals that have certain common characteristics that distinguish them from other groups within the same species, and that can be transmitted from generation to generation with a high degree of consistency. This definition of breed is somewhat similar to that given for species earlier except that it is more restrictive. The particular characteristics of a breed are determined by a breed association which has been organized by breeders to improve the breed. The association keeps a record of the ancestry of each animal that is registered with it. The record of an individual's ancestors is called a pedigree. Animals registered with a breed association usually must have parents that are also registered with the association. Such animals are usually referred to as purebred animals.

Breed associations may also keep performance records and other important information concerning the animals that are registered. Breeders who are members of the breed association pay a fee to belong to the association or to register their animals. These funds help pay for the expenses of the association which sometimes include the publishing of a book that lists the names and parents of all the registered animals. Other costs may include those associated with the promotion of the breed at fairs and livestock shows, the publishing of a breed magazine, and the employment of a general secretary.

BREED DEVELOPMENT

Occasionally, an individual or group of individuals will decide to develop a new breed because they desire a particular combination of characteristics that are not present in an existing breed. Among the first steps in the development of a new breed is to decide upon the particular combination of characteristics that the new breed should possess, and then to find the foundation animals that have one or more of those characteristics.

Animals that are to be used as foundation stock for the new breed may be chosen from among existing U.S. breeds or they may be imported from other countries. It is not easy to bring livestock into the United States from other countries because of strict import regulations. These regulations are imposed to protect the animals that are already here from diseases that could be brought in with the imported animals.

A procedure that has the same effect as importing sires from other countries is the shipping in of semen from other countries. Females of existing breeds can be inseminated with the imported semen to produce some crossbred foundation stock. Various crosses can be made using the crossbred stock, selected animals of existing breeds, and additional semen from foreign breeds. By using the imported semen to inseminate the offspring of each succeeding generation, a foreign or "exotic" breed can be established in this country in about four or five generations without ever importing one single animal of the breed.

Once a group of animals is put together that has the desired traits, a breeding program designed to increase their numbers and improve their quality is begun. For a time there may be periodic infusions of genetic material from outstanding individuals from outside the group. Eventually the breed is closed to outside sources of genetic material and all efforts are directed toward the improvement of the new breed.

Some examples of breeds that were developed from existing breeds are the Santa Gertrudis, Brangus, Beefmaster, and Charbray breeds of beef cattle, and the Columbia, Montedale, and Delaine breeds of sheep. A more detailed list of the various new breeds will be presented on the next several pages.

For the person who will be working in an agricultural profession, or in a field closely related to agriculture, it is important to be able to recognize the various breeds of livestock. The credibility of those who work with livestock producers is certainly enhanced if they are able to identify the breed of livestock belonging to the producer. It is also helpful to know enough about the breed to be able to carry on at least a brief conversation about it. It is therefore advantageous for aspiring young agriculturists to learn to recognize the various breeds of farm animals.

BREEDS AND TYPES OF CATTLE

Cattle are generally categorized into four groups based upon type. As used here, type refers to a group of animals within a species with the size and conformation best suited for a particular function. Conformation refers to the shape and form of the body. Cattle are usually classified into beef, dairy, dual-purpose, or draft types. In the United States, most cattle are of the beef or dairy type. A few dual-purpose cattle are used for beef and dairy purposes. Since cattle are generally not used for draft purposes in the United States, very few of that type are to be found in this country.

Breeds of Dairy Cattle

Most of the milk produced in the United States comes from the Holstein, Guernsey, Jersey, Ayrshire, Brown Swiss, and Milking Shorthorn breeds of dairy cattle (Fig. 2-1). The vast majority of this milk is produced by the Holstein breed. A number of comparative characteristics of these breeds is presented in Table 2-1. Note that the Holstein has the greatest average production of milk but the lowest percentage of butterfat. The Jersey breed produces the lowest average volume of milk but has the highest average butterfat percentage. Even though Holsteins generally produce milk with the

Holstein

(a)

Brown Swiss

(b)

Jersey

(c)

Figure 2-1   Breeds of dairy cattle. (a) Holstein (courtesy Robthom Farm, Springfield, Missouri). (b) Brown Swiss (courtesy Blessing Farms, Ft. Wayne, Indiana). (c) Jersey (courtesy Jim Chaney, Bowling Green, Kentucky).

Guernsey

(d)

ayrshire

(e)

Milking Shorthorn

(f)

Figure 2-1 Breeds of dairy cattle *continued*. (d) Guernsey (courtesy Boulder View Farm, Baraboo, Wisconsin). (e) Ayrshire (courtesy Stone House Farm, Orleans, Vermont). (f) Milking Shorthorn (courtesy Sancrest Farms, Billings, Missouri).

TABLE 2-1. BREEDS OF DAIRY CATTLE

Breed	Average Annual Milk Production (lb)	Average Butterfat Percentage	Approximate Number of Annual Registrations	Weight of Mature Cow (lb)	Country of Origin	Color
Holstein	14,700	3.6	280,000–330,000	1500	Holland	Black and white, some red and white
Jersey	9500	5.0	35,000–40,000	1000	Isle of Jersey	Fawn, some with white, some with blackish
Guernsey	10,300	4.6	25,000–30,000	1100	Isle of Guernsey	Fawn or yellow with white
Brown Swiss	12,700	4.0	13,000–17,000	1400	Switzerland	Dark to very light brown
Ayrshire	11,600	3.9	11,000–13,000	1200	Scotland	Red or mahogany with white
Milking Shorthorn	10,400	3.7	4000–5000	1250	England	Red, roan, or white

lowest percentage of butterfat, because of their higher milk production their total butterfat production is comparable to any of the
other breeds. In fact, the record for the most pounds of butterfat
produced in one year is held by a Holstein cow.

The average mature weight of cows is highest for the Holstein
breed and lowest for Jerseys. The other breeds are intermediate to
the Holstein and Jersey in varying degree with regard to milk production, fat percentage, and mature cow size. Milking Shorthorns are
generally thought of as dual-purpose cattle. Also included in Table
2-1 are the approximate numbers of cattle registered by each breed
each year. Even though most of the cattle within the several breeds
are not registered, these numbers are presented as indicators of the
relative popularity of the breeds.

Holsteins are probably the easiest of the dairy breeds to identify because of their distinctive black and white color pattern. Brown
Swiss cattle are almost as large as Holsteins at maturity, but due to
their usual light tan to grayish brown color, they are not likely to
be confused with Holsteins. Guernsey cattle are yellow and white
and are usually 300 to 400 pounds lighter than Holsteins and Brown
Swiss. Ayrshires are either red with white spotting or mahogany with
white spotting. Mahogany is a very deep red, almost maroon. The
coloration among Jersey cattle is probably the least consistent of the
several dairy breeds. The color may be fawn or buckskin, usually
without white spotting, although spotting sometimes is seen. In some
Jerseys, the color may be brown with shades of color varying from
brown to black on various parts of the body, particularly on the
head and neck. Jersey cattle very commonly have a considerable
amount of black on the face. Even though the pictures are not in
color, Figure 2-1 shows a number of the differences that exist among
the breeds of dairy cattle.

Breeds of Beef Cattle

Most of the beef in the United States is produced from Hereford and
Angus cattle, although a considerable amount is produced by dairy
cattle, particularly Holsteins. Earlier in this century, Shorthorn cattle
were a large contributor of beef, but their numbers have declined
considerably during the past 30 to 40 years.

Shorthorn cattle were imported to the United States from England almost 200 years ago. They are the oldest of today's breeds of beef cattle in this country. Some might argue that the Texas Longhorn is the oldest breed; however, the longhorn cattle of 200 years ago were of nondescript breeding with very little being done toward their improvement. They were descendants of cattle brought to this continent by Columbus and other early explorers. It has been only recently that the Texas Longhorn was developed and recognized as a breed.

The distinctive white-face pattern of the Hereford was first introduced to the United States over 160 years ago. Statesman and politician Henry Clay was responsible for having imported a few head to this country in 1817. The early Herefords were horned, but occasionally some natural polled calves would be produced. It was not until the end of the nineteenth century that an attempt was made to develop a line, and eventually a breed, of polled Hereford cattle. The development of polled Shorthorn cattle began about the same time.

Horned Herefords as a breed are now referred to as American Herefords. Since Hereford cattle were originally imported from England, the question arises as to why they should be called American Herefords. During the many years since the breed was first introduced to this country, the direction of their development has been toward the needs and desires of breeders in the United States. There is probably very little similarity between the way the breed has developed in the United States and the way is has developed in England. It is quite logical to think of U.S. Herefords now as an "American" breed. The same kind of statement could be made about the development of the Angus, Shorthorn, and Holstein breeds of cattle, as well as any other breed imported during the nineteenth and early twentieth centuries.

Angus cattle made their appearance in this country during the 1860s when they were brought from Scotland. They are a rather distinctive breed with their black coat and absence of horns. Cattle of the Angus and Hereford breeds account for the greatest proportion of the commercial and purebred beef herds in the United States. Breeds of beef cattle are described in Tables 2-2, 2-3, and 2-4. Pictures of twelve of these breeds are shown in Figure 2-2.

The Brahman breed of cattle was developed from several breeds of cattle brought from India in the mid-nineteenth century. As a group, the Indian breeds of cattle are also known as Zebu cattle.

TABLE 2-2.   EUROPEAN BREEDS OF BEEF CATTLE
COMMON IN THE UNITED STATES

Breed	Origin	Year Imported	Approximate Number of Annual Registrations	Characteristics
Angus	Scotland	1860	250,000–350,000	Black; polled
Red Angus	U.S.A. from Angus		10,000–15,000	Red; polled
Hereford	England	1817	200,000–250,000	Red with white face; horned
Polled Hereford	U.S.A. from Hereford		170,000–200,000	Red with white face; polled
Shorthorn	England	1783	18,000–26,000	Red, roan, or white; horned or polled
Charolais	France	1930	30,000–60,000	White; polled or horned
Simmental	Switzerland	1967	50,000–100,000	Red or yellow with white spots; horned
Limousin	France	1968	25,000–30,000	Red, rust, or wheat
Maine-Anjou	France	1970	9000–14,000	Dark red or mahogany with white spots
Chianina	Italy	1971	4000–6000	White, horned; largest of breeds

They are able to cope with hot tropical climates better than the European breeds because of the larger body surface area compared to their total weight, and also because of their ability to perspire. British and Continental European breeds of cattle have very few sweat glands. Zebu cattle are also more resistant to infestation by a number of parasites.

A number of newer breeds have been developed in part from Brahman foundation stock during the past half-century. Among these breeds are the Beefmaster, Santa Gertrudis, Brangus, Charbray, and Braford. About three-eighths of the genetic material in the Santa Gertrudis, Brangus, and Braford originated from Brahman. The other five-eighths came from the Shorthorn, Angus, and Hereford breeds. Charbray is composed of from one-eighth to one-fourth Brahman

TABLE 2-3. BREEDS OF BEEF CATTLE IN THE UNITED STATES DEVELOPED IN PART FROM ZEBU CATTLE

Breed	Source or Origin	Year When Development Began	Approximate Number of Annual Registrations	Characteristics
Brahman	Three Zebu breeds: Nellore, Guzerat and Gir.	Late 1800s	25,000–35,000	Gray or red; prominent hump; drooping ears; pendulous skin under throat and dewlap; horned
Beefmaster	Lasater Ranch, Texas. $\frac{1}{2}$ Braham, $\frac{1}{4}$ Shorthorn, $\frac{1}{4}$ Hereford.	1908	15,000–25,000	Several colors but usually red; usually horned
Santa Gertrudis	King Ranch, Texas. $\frac{3}{8}$ Braham, $\frac{5}{8}$ Shorthorn.	1920	25,000–30,000	Cherry red; usually horned; Brahman features but not as pronounced
Charbray	Southwest Texas. $\frac{1}{8}$ to $\frac{1}{4}$ Brahman, $\frac{3}{4}$ to $\frac{7}{8}$ Charolais.	Late 1930s	Registered with Charolais	White
Brangus	Clear Creek Ranch Welch, Oklahoma. $\frac{3}{8}$ Braham, $\frac{5}{8}$ Angus.	1942	12,000–16,000	Black; polled
Barzona	Bard Ranches, Arizona. Afrikander, Hereford, Santa Gertrudis, and Angus.	1942	Fewer than 1000	Red
Braford	Adams Ranches, Florida. $\frac{3}{8}$ Braham, $\frac{5}{8}$ Hereford.	1948	Fewer than 1000	Hereford markings, some are brindled; horned
Ranger	Wyoming and California. From several breeds.	1950	Fewer than 1000	Numerous colors

TABLE 2-4. MISCELLANEOUS BREEDS OF CATTLE IN THE UNITED STATES

Breed	Origin	Characteristics
Beef Friesian	U.S.A.	Developed from dual-purpose European Friesians
Belted Galloway	Scotland	Black with white belt encircling the body between shoulders and flanks; polled
Blonde d'Aquitaine	France	Various shades of yellow
Devon	England	Ruby red; white horns
Fleckvieh	Germany	Red and white; horned; evolved from Simmental
Galloway	Scotland	Black, some belted; polled
Gelbvieh	Germany	Light red; horned; imported in 1972
Hays Converter	Canada	Black with white face; developed by Harry Hays, former minister of Agriculture beginning in 1957
Dutch Belt	Holland	Black with white belt similar to Belted Galloway
Marchigiana	Italy	Grayish white; small horns
Murray Grey	Australia	Silver gray; polled; from Angus and Shorthorn
Normande	France	Dark red speckled on white
Norwegian Red	Norway	Red, some have white; horned; dual-purpose
Pinzgauer	Austria	Brown with white top line extending over tail region and down under body; sides colored
Red Poll	England	Red; polled; dual-purpose
Salers	France	Cherry red, white switch; usually horned
Scotch Highland	Scotland	Brown, fawn, or yellow; long horns; long hair
South Devon	England	Light red; dual-purpose
Sussex	England	Mahogany red
Tarentaise	France	Light brown to blond
Texas Longhorn	U.S.A.	Variegated colors; long horns; developed from cattle brought over by Columbus in 1493
Welsh Black	Wales	Black; horned

(a)

(b)

(c)

Figure 2-2 Some common breeds of beef cattle. (a) Angus. (b) Simmental. (c) Hereford. (Courtesy Curtiss Breeding Industries Inc., Elburn, Illinois)

(d)

(e)

(f)

Figure 2-2 Some common breeds of beef cattle *continued.* (d) Maine-Anjou. (e) Charolais. (f) Chianina. (Courtesy Curtiss Breeding Industries Inc., Elburn, Illinois)

46

(g)

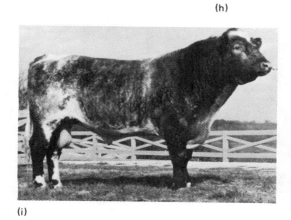

(h)

(i)

Figure 2-2 Some common breeds of beef cattle *continued*.
(g) Brahman (courtesy Curtiss Breeding Industries Inc., Elburn,
Illinois). (h) Santa Gertrudis (courtesy American Breeders Service,
DeForest, Wisconsin). (i) Shorthorn (courtesy Lazy D Ranch, Ex-
celsior Springs, Missouri and Oakhill Farm, Austin, Minnesota).

(j)

(k)

(l)

Figure 2-2 Some common breeds of beef cattle *continued*. (j) Brangus (courtesy Cedar Springs Ranch, Greenbrier, Arkansas). (k) Pinzgauer (courtesy Tal Bauernhof, Inc., Lonsdale, Minnesota). (l) Limousin (courtesy American Breeders Service, DeForest, Wisconsin).

and the rest Charolais. The Beefmaster breed was developed with about 50 percent of its genes originating from the Brahman, 25 percent from the Hereford, and the other 25 percent from the Shorthorn.

The first Charolais cattle imported into the United States came by way of Mexico in the 1930s. Many of the bulls were bred to existing breeds, particularly white Shorthorns. Succeeding generations were bred back to Charolais bulls until, after five generations, the proportion of Charolais breeding reached 31/32. At this point they were registered as purebreds.

Since the introduction of the Charolais breed, many breeds from the continent of Europe have been introduced. Because of problems with hoof-and-mouth disease, as well as other diseases, importations are very strictly regulated. The breeds that have been imported during the period since the coming of the Charolais are called exotic breeds. The term exotic means strange, foreign, or not native. However, after a period of time, exotic things cease to be strange and unfamiliar and thus are no longer exotic. This is what has happened to the Charolais; it is generally no longer thought of as an exotic breed.

The importance of the exotic breeds lies in the fact that they are a source of new genes that may be useful in the improvement of the beef cattle industry. Only time will tell how important a role any one breed will play. Some breeds will become established as has the Charolais, while others will never play any significant role. The more common of the exotic breeds at present, in addition to the Charolais, are the Simmental, Maine-Anjou, Limousin, and Chianina (pronounced KEY-a-NEEN-a). They are described in Table 2-2.

A number of new breeds are now being developed from crosses between Brahman and Simmental, Brahman and Limousin, and Angus and Chianina, as well as from other combinations. It is anticipated that several more new breeds of cattle will be developed during the next few years.

BREEDS OF SWINE

The two most popular breeds of swine according to recent registration figures are the Durocs and the Yorkshires. Hampshires have been, and still are, a very popular breed, too. For a number of years these three breeds have been represented in the vast majority of pure-bred and commercial herds in the United States. The number of registered animals in the Spotted breed has been increasing quite rapidly during the past ten years (see Table 2-5). Registration figures for Landrace hogs have also taken a recent dramatic upswing after this breed had experienced a number of years of low popularity.

There has been a significant increase in the number of all registered hogs in the past five or six years. In 1979, almost 1 million hogs were registered in the United States. If we assume that this number represents about one-third of all registered hogs, there would be approximately 3 million registered swine. With about 55 million total hogs in the country, the registered animals represent between 5 and 6 percent of the total number of hogs. Although most of the pork in the United States is produced from crossbred hogs, pure-bred herds play a very important role in the swine industry. They are the source of breeding stock for other purebred herds and, par-

TABLE 2-5.　COMPARISON OF NUMBERS OF REGISTERED SWINE AMONG THE COMMON BREEDS IN THE UNITED STATES

Breed	1970	1973	1976	1979
Duroc	76,394	71,413	155,793	258,245
Yorkshire	50,506	43,433	165,740	246,356
Hampshire	74,101	51,135	155,817	143,716
Spotted	13,974	17,471	98,616	130,023
Landrace	8810	6518	10,286	73,312
Chester White	19,934	18,992	40,796	62,680
Poland China	16,102	7983	13,778	18,905
Berkshire	8012	5144	12,777	15,012
Total of Eight Breeds	267,833	222,111	653,603	948,249
Total Number of Swine (000)	67,540	61,106	55,085	59,860

**TABLE 2-6.   COMMON BREEDS OF SWINE
IN THE UNITED STATES**

Breed	Country or State of Origin	Color	Distinguishing Features
Duroc	New York and New Jersey	Red	Drooping ears
Yorkshire	England	White	Erect ears; dished face
Hampshire	Kentucky	Black with white belt	Erect ears
Spotted	Indiana	Large black spots on white	Drooping ears
Landrace	Denmark	White	Long drooping ears; long body
Chester White	Pennsylvania	White	Drooping ears
Poland China	Ohio	Black with some white spots	Drooping ears
Berkshire	England	Black with white points	Erect ears; dished face

ticularly in the case of purebred boars, are a major source of superior breeding stock for commercial herds. Because of this important relationship between the purebred and commercial segments of the swine industry, the number of registered swine provides a reasonably good index of the relative popularity of the various breeds.

Characteristics of the eight common breeds of swine are presented in Table 2-6. Photographs of these breeds are shown in Figure 2-3. Learning to recognize the breeds of swine is probably easier than for any other class of farm animal. Three of the breeds are all white. Among them the Yorkshire is the easiest to identify because of its erect ears and prominent "dished" face. The Chester White and Landrace breeds both have drooping ears, but the Landrace has longer ears and a longer body. In fact, hogs of this breed have one more vertebra than the other breeds have.

Among the four breeds that are black and white, the Hampshire and Spotted breeds are quite distinctive. The white of Hampshires is

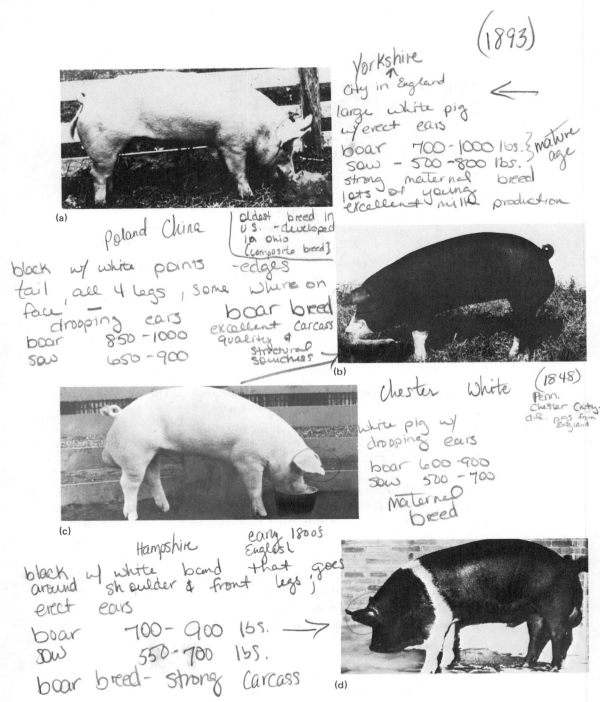

(1893)

Yorkshire
↑
City in England
large white pig
w/ erect ears
boar   700-1000 lbs. } mature age
sow - 500-800 lbs.
strong maternal breed
lots of young
excellent milk production

Poland China
oldest breed in U.S. - developed in ohio (composite breed)
black w/ white points -edges
tail, all 4 legs, some where on
face -
    drooping ears   boar breed
boar   850-1000   excellent carcass
sow   650-900   quality &
                                    structural
                                    soundness

Chester White   (1848)
Penn. Chester County.
diff. pigs from England
white pig w/
drooping ears
boar 600-900
sow   500-700
Maternal
    breed

Hampshire   early 1800's
                        Englash
black w/ white band that goes
around shoulder & front legs;
erect ears
boar   700-900 lbs. →
sow   550-700 lbs.
boar breed- strong carcass

Figure 2-3   The common breeds of swine. (a) Yorkshire. (b) Poland China. (c) Chester White. (d) Hampshire. (Courtesy University of Illinois Vocational Agriculture Service, Urbana, Illinois)

52

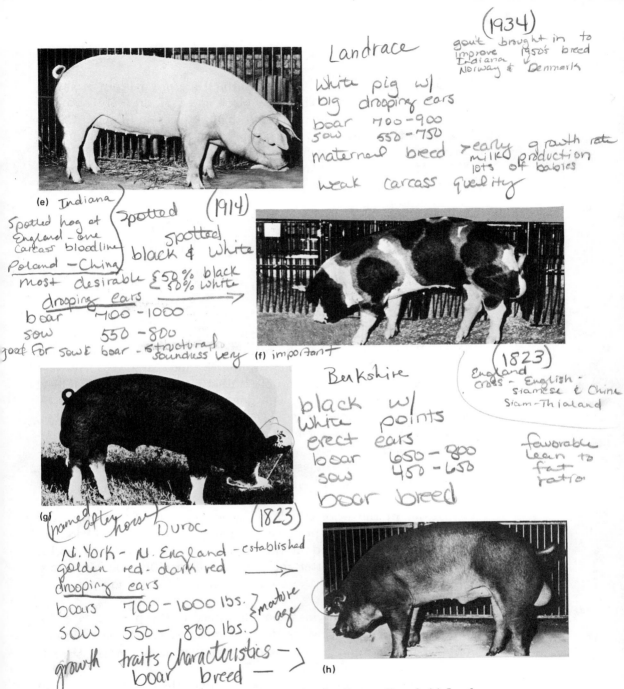

Landrace (1934)

goat brought in to improve 1950's breed

Indiana

Norway & Denmark

White pig w/ big drooping ears

boar 700-900
sow 550-750

maternal breed → early growth rate
milk production
lots of babies

weak carcass quality

(e) Indiana

Spotted hog of England - one carcass bloodline Poland - China

Spotted (1914)

spotted black & white

most desirable { 50% black / 50% white →

drooping ears

boar 700-1000
sow 550-800

goat for sow & boar - structural soundness very (f) important

Berkshire (1823)

England cross - English - Siamese & Chine Siam - Thialand

black w/ white points
erect ears

boar 650-800
sow 450-650

boar breed

favorable lean to fat ratio

(g) named after horse **Duroc** (1823)

N. York - N. England - established
golden red - dark red →
drooping ears

boars 700-1000 lbs.   } mature age
sow 550 - 800 lbs.   }

growth traits characteristics →
boar breed —

(h)

Figure 2-3   The common breeds of swine *continued*. (e) Landrace. (f) Spotted. (g) Berkshire. (h) Duroc. (Courtesy University of Illinois Vocational Agriculture Service, Urbana, Illinois)

as unique among swine breeds as the Hereford pattern is among breeds of beef cattle. The large black spots of the Spotted breed allow it to be readily identified. Both Berkshires and Poland Chinas are black with white feet and nose. However, the white markings of the Poland China show more variation than do those of the Berkshire. The features that make these breeds easy to distinguish are the erect ears and dished face of the Berkshire. The ears of the Poland China droop.

The Duroc is the only red breed of the eight, and also the only one that shows no white.

BREEDS OF SHEEP

Sheep in the United States are raised primarily for their wool and meat. Breeds used primarily for their wool include Rambouillet, Merino, Debouillet, Corriedale, Columbia, and Targhee. They were developed for their quality wool, mothering ability, milking ability, and reproductive efficiency. The finest quality wool is produced by the Rambouillet, Merino, and Debouillet breeds (see Chapter 14). Breeds used primarily for meat purposes include the Suffolk, Hampshire, Shropshire, Oxford, Southdown, Dorset, Cheviot, and Montedale. Wool is important as a secondary product from these breeds. The meat breeds are often referred to as medium wool breeds of sheep.

Ewes of the wool breeds are generally crossed to rams of the larger meat breeds to produce crossbred market lambs. Crossbred ewes are often involved in the commercial production of lambs. Crossbred ewes are better mothers, more prolific, and give more milk. Characteristics of the common breeds of sheep are presented in Table 2-7.

The Suffolk is the most popular breed of sheep as indicated by the number of annual registrations. Hampshires are the second most popular breed. Among the next most popular breeds are the Dorset, Rambouillet, Columbia, and Corriedale.

To the inexperienced observer, many of the breeds of sheep are difficult to distinguish one from the other. However, several of the breeds have certain distinctive features that are quite helpful in the identification of those breeds. Some of these features can be seen

TABLE 2-7.   BREEDS OF SHEEP
IN THE UNITED STATES

Breed	Country of Origin	Body Size	Fleece	Characteristics
Rambouillet	France	Large	Fine	Horned rams
Merino	Spain	Large	Fine	Horned rams
Debouillet	U.S.A.	Medium	Fine	Horned rams
Targhee	U.S.A.	Medium large	Medium fine	Polled
Corriedale	New Zealand	Medium large	Medium	Polled
Columbia	U.S.A.	Large	Medium	Polled
Suffolk	England	Large	Medium	Polled; black head; head free of wool
Hampshire	England	Large	Medium	Polled; black face
Shropshire	England	Medium	Medium	Polled; black face
Oxford	England	Large	Medium	Polled; black nose
Cheviot	Scotland	Small	Medium	Polled; erect ears
Southdown	England	Small	Medium	Polled; wool extends below eyes
Montedale	U.S.A.	Medium	Medium	Polled; head free of wool
Finnsheep	Finland	Small	Medium	Polled; head free of wool
Dorset	England	Medium	Medium coarse	Polled or horned
Romney	England	Medium	Coarse	Polled

in the photographs in Figure 2-4. Black is present in varying degrees on the heads and legs of Suffolk, Hampshire, Shropshire, and Oxford breeds. Suffolks express the greatest amount of black with Oxfords showing the least amount. Hampshires and Shropshires show an intermediate amount of black; however, Shropshires are usually smaller in size and have more wool on their heads than Hampshires.

Distinctive features of the Cheviot include the erect ears and clean head and legs. The horns of Rambouillet rams are certainly distinctive features. Dorset rams often have horns, too, but the fine wool of the Rambouillet and the medium coarse wool of the Dorset are helpful in the distinguishing between these two breeds. Although

Merino ram

Rambouillet ram

Cheviot ram

Columbia ram

Corriedale ram

Dorset ram

Hampshire ram

Oxford ram

Shropshire ram

Southdown ram

Suffolk ram

the wool of the Columbia is not as fine as that of the Rambouillet, it is much finer than that of the Dorset. The Southdown is rather distinctive because of its small size.

BREEDS AND TYPES OF POULTRY

Commercially, chickens are categorized as to whether they are egg-type or meat-type birds. The White Leghorn was developed as an egg-laying type. It is a medium-sized breed that lacks the meatiness of the meat-type breeds. The meat breeds include the Plymouth Rock, New Hampshire, Rhode Island Red, and Cornish. No breed associations exist for individual breeds as they do for sheep, cattle, horses, and swine. The *American Standard of Perfection* serves as a guide to the characteristics of the so-called purebred chickens. This document is published by the American Poultry Association. Some common breeds are described in Table 2-8 and shown in Figure 2-5.

Within the standard breeds of chickens, various strains of birds have been developed. A pure strain is one where no outside source of breeding has been introduced for several generations. Two different pure strains within a breed may be crossed to produce a strain cross. Breed crosses are produced by crossing two different strains from different breeds. Inbred hybrids are produced from two different pure strains of the same breed. They differ from strain crosses in that their parent lines are much more highly inbred than are the pure

TABLE 2-8.   BREEDS OF POULTRY

Breed	Origin	Plumage	Egg Color	Use
White Leghorn	Italy	White	White	Egg production
Plymouth Rock	U.S.A.	White or barred	Brown	Crossbreeding for meat
Rhode Island Red	U.S.A.	Deep red	Brown	Meat production
New Hampshire	U.S.A.	Rusty red	Brown	Meat production
Cornish	England	White or dark	Brown	Crossbreeding for meat
Brahma	Asia	Columbian	Brown	Show, hobby, meat
Cochin	Asia	Buff	Brown	Show, hobby

Figure 2-5 Some common breeds of chickens. (a) White Leghorn. (b) Cornish. (c) White Plymouth Rock. (d) Barred Plymouth Rock. (e) New Hampshire. (f) Rhode Island Red. (United States Department of Agriculture)

strains. The nature and functions of inbreeding are presented in Chapter 7. The several breeding methods just described have been used extensively in the production of today's egg-laying and broiler chickens.

Most of the eggs produced commercially in this country are from various strains and combinations of strains of the White Leghorn breed. Broilers are produced from various combinations and crosses involving the White Plymouth Rock and the White Cornish breeds. Most of these chickens are produced by various commercial firms whose breeding practices are trade secrets.

Turkeys are native to North America. The turkeys of today are descendants of those native turkeys that explorers as far back as the days of Columbus found when they came to this continent. Breeding and selection have resulted in a much more meaty bird, particularly in the breast region. While selection for meatiness has been quite successful, it has resulted in the production of clumsier males which makes it very difficult for them to perform the natural mating act. This is one of the reasons why artificial insemination has come to be used rather extensively in turkeys. Most turkeys raised commercially today are of the Broad Breasted Large White breed.

BREEDS OF HORSES

Numbers of the various types and breeds of horses have undergone considerable changes since the turn of the century. The numbers of horses and mules reached a high of around 26 million in 1915, 5 million of which were mules. Draft horses and mules provided the power for farming; stock horses helped in the working of cattle; and horses pulled carriages, buggies, and wagons in the cities as well as on the farms. Numbers of horses and mules then followed a steady decline until, by 1960, there were only about 3 million such animals in the country. This steady decline was associated with the development and increase in the numbers of farm tractors, power machinery, and trucks. The stock horse was about the only type of horse that still had a job to do. Since 1960, the number of horses has been increasing until there are now in excess of 10 million horses in the country. Characteristics of several breeds are presented in Table 2-9. Pictures of several common breeds are shown in Figure 2-6.

TABLE 2-9. COMMON BREEDS OF HORSES IN THE UNITED STATES

Breed	Height (hands)	Weight of Mature Mare (lb)	Origin	Primary Use and Other Characteristics
American Quarter Horse	$14\frac{1}{2}-15\frac{1}{2}$	1000–1250	U.S.A.	Stock, racing; well-muscled
Thoroughbred	$15\frac{1}{2}-17$	1000–1300	England	Racing, polo, hunters
Appaloosa	$14\frac{1}{2}-15\frac{1}{2}$	1000–1250	Northwest U.S.A.	Stock, pleasure; dark spots on white over hips, often spotted over whole body
Arabian	$14\frac{1}{2}-15\frac{1}{2}$	850–1000	Arabia	Pleasure, show; short body
Standardbred	$15\frac{1}{2}-17$	1000–1300	U.S.A.	Harness racing, harness shows
Tennessee Walking Horse	15–16	1000–1200	Tennessee	Pleasure, Walking Horse shows
Paint	$14-15\frac{1}{2}$	900–1100	U.S.A.	Stock, pleasure; white on any color
American Saddle Horse	15–16	1000–1200	Kentucky	Pleasure, 3- and 5-gaited shows
Morgan	$14\frac{1}{2}-15\frac{1}{2}$	1000–1200	New England	Pleasure, stock
Pinto	$14-15\frac{1}{2}$	900–1100	U.S.A.	Stock, pleasure; half white with one or two other colors, no glass eye (white eye).
Missouri Foxtrotting Horse	$14\frac{1}{2}-15\frac{1}{2}$	900–1100	Ozarks	Pleasure, stock, trail riding
Shetland Pony	$9\frac{1}{2}-11\frac{1}{2}$	300–400	Shetland Isles	Children's mount, harness pony
Hackney Pony	$11\frac{1}{2}-14\frac{1}{2}$	450–850	England	Fine harness and carriage pony
Belgian	$15\frac{1}{2}-17$	1900–2400	Belgium	Most "drafty" of the breeds
Percheron	$15\frac{1}{2}-17$	1600–2200	France	Black or dapple gray
Clydesdale	$15\frac{1}{2}-17$	1700–2000	Scotland	Tall, good action, white legs

(b)

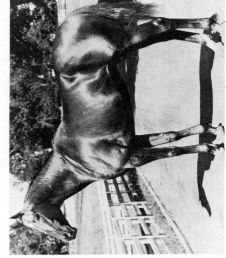

(d)

(a)

(c)

Figure 2-6 Some common breeds of horses. (a) Arabian. (b) Appaloosa. (c) American Saddle Horse. (d) Thoroughbred. (United States Department of Agriculture)

(e)

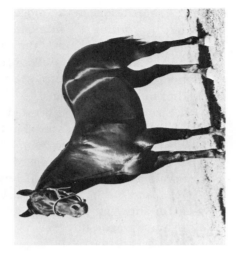

(f)

(g)

(h)

Figure 2-6 Some common breeds of horses *continued.* (e) Tennessee Walking Horse. (f) Standardbred. (g) Morgan. (h) American Quarter Horse. (United States Department of Agriculture)

Today, most breeds of horses are used for pleasure riding, racing, and for various kinds of shows. Stock horses are still used to work cattle on farms and ranches. The American Quarter Horse is such a breed, and is the most popular breed of horse in the United States. According to numbers of annual registrations, next to the American Quarter Horse, the most popular breeds of horses in the United States are the Thoroughbred, Appaloosa, Arabian, and Standardbred. The Thoroughbred and the Standardbred are breeds used primarily for racing.

Size in horses is expressed in terms of height at the withers and of weight. Height is measured in hands; a hand is equal to four inches. Horses that are shorter than $14\frac{1}{2}$ hands are generally called ponies. The Shetland is an example of a pony breed. At the other end of the size scale are the draft breeds, breeds that were developed for pulling wagons and farm machinery. Today, most of the draft horses are used for show purposes, in pulling contests, and in teams of six- and eight-horse wagon hitches.

Most breeds of light horses, those in the size range between ponies and draft horses, can be identified by their distinctive gaits. The gait of a horse refers to the action of the feet and legs when it runs and walks. Among these distinctive gaits are the trot and pace of the Standardbred, the running walk of the Tennessee Walking Horse, the rack of the American Saddle Horse, and the foxtrot of the foxtrotting breeds. Many breeds perform the trot and canter, but each breed performs them in a way that is rather unique to that breed.

For most breeds, color is not an important criterion for identification, although certain colors are quite often more popular than others. Some of the common colors include black, brown, and the reddish color of the chestnut or sorrel. The bay is rather distinctive with its brown or reddish brown body and black mane, tail, and legs. The buckskin has the same black markings on the legs, mane, and tail, but the body is tan or fawn colored. The grulla also has the black markings of the bay, but the body is mouse colored. Many variations of white markings on the body, face, legs, and the mane and tail can be observed.

Some breeds have been developed using a certain color or color pattern as a trademark. Among these so-called color breeds are the Appaloosa, Paint, Pinto, and Albino. Palomino and buckskin are popular colors, and registry associations have been established for horses that express these colors. However, they cannot be correctly considered as breeds because of the way in which the colors are inherited. For example, when palomino horses are mated together, only about one-half of the foals produced will be palomino in color. Because of this, it is common to say that these colors do not "breed true." The inheritance of the palomino color is explained in Chapter 5.

STUDY QUESTIONS AND EXERCISES

1. Define each of the following terms:

   breed   purebred
   conformation   pure strain
   draft   strain
   exotic   strain cross
   gait   Zebu
   pedigree

2. Compare the meanings of the terms species and breed.

3. What is the purpose of a breed registry association?

4. Describe two ways to establish a new breed.

5. What is the most popular breed of dairy cattle in the United States?

6. Name the three leading breeds of beef cattle in the country.

7. Compare the average volume of milk produced annually and percentage of butterfat for the common breeds of dairy cattle.

8. Name the country of origin for each breed of farm mammal.

9. Compare the foundation sources of the several breeds of cattle that were developed in part from Brahman.

10. What are some factors that could determine which new breeds of cattle will become established and which may become of lesser importance?

11. What are the three most popular breeds of swine in the United States?

12. About what percentage of all swine in the country are registered?

13. What kind of breeding program is used to produce most of the hogs in the United States?

14. Which of the common breeds of swine are white? Which are red? Which have drooping ears? Which originated in England?

15. Describe some differences in the wool breeds and meat breeds of sheep.

16. Which of the common breeds of sheep are fine wool breeds? Which have black on the face? Which developed in the United States?

17. Which is the most common breed of egg-producing chickens?

18. What is the origin of most of today's broiler chickens?

19. Why has artificial insemination become important in the commercial production of turkeys?

20. Trace the changes in the numbers of horses during the twentieth century.

21. Name the five most popular breeds of horses in the United States.

22. Identify each breed of horse as to its primary use and country or region of origin.

3

Anatomy and Physiology

One of the first steps in learning about any new subject is to learn its language. You will have to learn many new terms as you study anatomy and physiology. Only a very few of them will be presented at the beginning of this chapter. By the time you have completed your study of the chapter, however, you will have learned many more. And as you read further in this book you will discover that the subject of anatomy and physiology is not limited to this one chapter. In particular, the anatomy and physiology of the male and female reproductive systems and of the digestive system are presented in some detail in later chapters. Because reproduction and nutrition play such an important role in the production of domestic animals, we will devote much more attention to these subjects than to the other systems of the body. It is the purpose of this chapter to present a brief overview of the subjects of anatomy and physiology. It begins with cells, tissues, and organs, and progresses to the various body systems. In particular, we will deal with the skeletal system; the nerves and muscles; the circulatory, respiratory, and urinary systems; and the endocrine system, including the most important hormones and their functions.

Anatomy deals with the form and structure of an organism. Physiology deals with the functions of various parts of the animal body from the single cell to the entire body as a unit. In order to understand how a particular part of the body functions, it is necessary to have a certain level of knowledge concerning the anatomy of that part and of the other parts of the body with which it is related. Anatomy and physiology go hand in hand. In order to make the study of these two sciences easier, they can be divided into a number of areas or categories. Gross anatomy, for example, deals with the structures of the body that can be seen and studied with the unaided eye. On the other hand, microscopic anatomy, or histology, involves the study of cells and tissues that can only be seen with the help of a microscope.

Systematic anatomy deals with the study of the body by systems. A system involves a group of organs working together to carry out a common overall function. For example, the circulatory system is made up of several organs that work together to transport the blood throughout the body. Each specialized field of systematic anatomy is usually referred to by a scientific name based on the Latin word describing the system that is being studied. Table 3-1 lists the major fields of systematic anatomy, the system that each focusses on, and the major structures associated with that system.

Comparative anatomy may deal with any particular structure or group of structures but will involve several species in the study. For example, the structure of the heart may be studied and compared in several species of vertebrates. Embryology deals with the development of the embryo and fetus of a mammal or the embryo of a chick or other vertebrate from fertilization until parturition or hatching. Cytology deals with the form and structure of plant and animal cells. Even though the various terms presented in the last few paragraphs have dealt specifically with anatomy, physiology can be included. Someone who is studying neurology for example, is not necessarily limited to studying only the anatomy of the nervous system. He or she may have an equal or even greater interest in the physiology of the system or one of its parts.

TABLE 3-1.   MAJOR FIELDS
                 OF SYSTEMATIC ANATOMY

Scientific Name	System	Major Structures
Angiology	Circulatory system	Heart and blood vessels
Arthrology	Articular system	Joints of the skeleton
Dermatology	Integumentary system	Skin
Endocrinology	Endocrine system	Hormone-producing glands
Myology	Muscular system	Muscles
Neurology	Nervous system	Brain, spinal cord, and nerves
Osteology	Skeletal system	Bones
Splanchnology	Digestive, reproductive, respiratory, and urinary systems	Essentially all the internal organs
Urology	Urinary system	Kidneys and bladder

Terms of Location and Direction

In describing certain parts of the body it is often necessary to use
terms that help to locate those parts. While terms such as left and
right, front and rear, and up and down may be useful, there are times
when they may not be descriptive enough. A number of terms are
used in the study of the animal body to denote direction and location. The term anterior indicates that something is located toward
the animal's head. The term cranial means the same thing. In the
study of the human body, the term superior means the same as
cranial and anterior. Each of these directional terms have antonyms,
or words that mean the opposite. Caudal is the opposite of cranial;
it means toward the tail. Posterior is the antonym of anterior.
Referring to humans, the term inferior is the opposite of superior. In
a cow, the pin bones are located caudal or posterior to the hook
bones. In humans, the ribs are cranial or superior to the pelvis.

Dorsal is a directional term indicating that something is located
toward or beyond the backbone or vertebral column. A saddle is
placed on the dorsal part of a horse. The term which means the
opposite of dorsal is ventral. The nipples are located on the ventral

part of a sow. The naval is located ventrally on all mammals. The terms proximal and distal are also antonyms. Proximal means that something is located near the point of attachment. It has the same root meaning as proximity which means nearness. Distal comes from the same root as distant. Distal indicates that something is located away from the point of attachment. The elbow is located proximal to the wrist while the fingers are located distal to the elbow.

The medial plane of an animal represents a cut through the vertebral column perpendicular to the surface of the earth when the animal is standing upright. In the slaughtering process, the carcass of a steer or barrow is usually cut into right and left halves through the medial plane. The term medial is a directional term which means toward the medial plane. Lateral indicates that something is located away from the medial plane. A sagittal plane would be any section through an animal parallel to, but not through, the medial plane. A transverse plane is one that is at right angles to the medial plane. A transverse cut would divide the body into a front or anterior half and a rear or posterior half. A T-bone steak represents a transverse slice of the loin and vertebral column of a steer.

Organization of the Animal Body

The building blocks of higher organisms are the cells. Cells are designed in such a way that most of them have many things in common. With very few exceptions, they all contain a nucleus. Inside the nucleus are found the genes and chromosomes which are the control center of the cell. Different types of cells have different functions. The overall function of a cell is usually related to materials located in the cytoplasm. Muscle cells are specially designed to contract due to the presence of a special substance located in their cytoplasm. Nerve cells carry nerve impulses because of the special characteristics of the nerve cell cytoplasm. Generally, the cytoplasm in any type of cell consists of everything that is outside the nucleus and inside the cell membrane.

Cells of similar structure and function are grouped together in the animal to form tissues. The body is composed of several types of tissue. Muscle cells would not be able to provide much contractile force alone but in the form of muscle tissue they are capable of moving the whole body. Other cells specialize in holding things together; they form connective tissue. Cells that specialize in transmitting nerve impulses group together to form nerve tissue. Epithelial tissue is made up of special cells designed to form protective coverings for other tissue. The skin is a good example of epithelial tissue. A fifth type of tissue is called fluid tissue; the blood is a good example. Blood is made up of special cells, the red blood cells and white blood cells, supported by a fluid medium called plasma. The blood has many functions including the transportation of nutrients and waste products.

The next level in the structural hierarchy of the animal body is that of the organs. Organs are composed of two or more types of tissue. They are designed to carry out functions too complex to be accomplished by a single type of tissue. The small intestine is an organ designed to carry out the digestion of several types of food. It is made up of connective tissue, muscle tissue, and epithelial tissue. Several organs working together to accomplish a common function are referred to as a system. The stomach, the small intestine, the large intestine, the pancreas, and the liver are organs that together form the digestive system. The various systems of the body were listed in Table 3-1.

Finally, all the systems, organized in just the right manner and interacting with each other, form the whole body. None of the cells, tissues, organs, or systems could operate independently, but as parts of the animal body they are able to do amazing things. All animals in the vertebrate group tend to be organized in a similar manner. The body takes on a certain shape or form basically due to the structure of the bony or cartilaginous skeleton. The form or shape of the body takes on an even more definite appearance due to the muscles and various types of connective tissue that cover the skeleton. The whole body is then covered with a protective layer of epithelium, called the skin, which includes special modifications called hair, feathers, or scales.

Several cavities are formed inside the body to hold the various other organs and systems. In mammals, the thoracic cavity holds the lungs and heart. The thoracic cavity is surrounded generally by the rib cage and by the diaphragm at the posterior end. The abdominal cavity is the largest cavity in the mammal, extending from the diaphragm to the pelvis and from vertebral column dorsally to the vertral wall of the body. Inside the abdominal cavity are found much of the digestive system and parts of the urinary and reproductive systems. Since birds and other nonmammalian vertebrates do not possess a diaphragm, in these animals the thoracic and abdominal cavities form one large body cavity. The pelvic cavity of vertebrates contains parts of the digestive, reproductive, and urinary systems. The cranial and spinal cavities contain the part of the nervous system called the central nervous system, which includes the brain and spinal cord.

The body is equipped with appendages referred to by various terms depending upon the species. They may be called legs, wings, fins, arms, or flippers. Their primary function is locomotion, although certain other tasks may be accomplished by these appendages.

Epithelial Tissues

Tissues that form protective coverings for other tissues are called epithelial tissues. These protective coverings are found on the inside of tubular or hollow organs as well as on the outside. A transverse section of the digestive tract shows five distinct layers of tissue: an outer layer of epithelium, a second layer of connective tissue, a third layer of smooth muscle, a fourth layer of connective tissue, and finally an internal layer of epithelium. Two types of epithelium are represented in this example. The outer layer of epithelium is referred to as a serous membrane or serosa. It secretes a mucous material to keep the organ wall moist and lubricated to resist friction. The mucus also prevents adhesion to other organs or to the wall of the body cavity. The inner layer of the digestive tract is made up of a type of epithelium some of which is specialized in the secretion of digestive enzymes. This type of epithelium is called glandular epithelium.

Glands are classified as either exocrine or endocrine. Exocrine glands release their secretory products through ducts to a specified location. Examples of exocrine glands include the salivary glands; the accessory glands of the male reproductive system; enzyme-secreting glands of the pancreas, stomach, and small intestine; and mucus-secreting glandular epithelium. Glands are specialized modifications of epithelial tissue and show varying degrees of development from simple to very complex. Some glands have developed to the point where there is no duct system to carry the secretion. These types of glands are called endocrine glands or ductless glands. Their secretions are called hormones and are transported throughout the body by the blood stream.

The skin is another kind of epithelium. It consists mainly of two layers, an outer covering of protective cells called epidermis, and an inner layer of dermis. The dermis is interspersed with connective tissue, nerves, and blood vessels. In mammals, the hair follicles are found in the dermis. In animals that perspire, sweat glands are found in the dermis. Figure 3-1 shows a cross section of the skin of a mammal. The skin is usually connected to the underlying tissues by a layer of loose connective tissue referred to as the hypodermis.

Several examples of modified skin or external epithelium can be given. The hair follicle develops from the epidermis. The color of

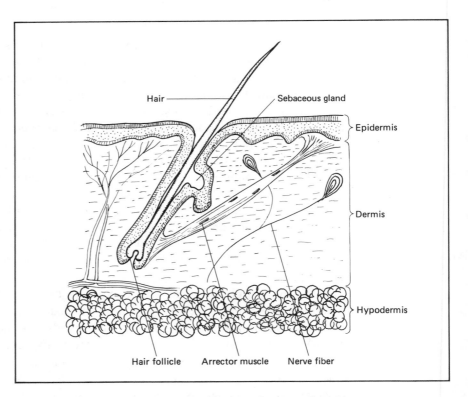

Figure 3-1   Cross section of mammalian skin.

hair is due to the distribution and location of a single pigment called melanin. The hair follicles are connected to a small muscle that causes the hair to become erect when the muscle contracts. The material that makes up the horns and hoof walls is also modified epithelial tissue. It is produced by special cells of the skin next to the base of the horn or hoof. The material making up the hoof or horn grows from the base and pushes the old growth outward. The hair grows out from inside the hair follicle in much the same way. The finger and toenails of humans also develop in very much the same way.

Connective Tissues

As the name implies, connective tissues function in binding other tissues together. They may provide shape, strength, or protection to certain organs. A kind of connective tissue called elastin is found in the walls of arteries, in the wall of the abdomen, and in the connective tissue that helps support the head. Both tendons and ligaments are made of a kind of connective tissue called collagen. The tendons connect muscle to bone while the ligaments connect bone to bone. Areolar connective tissue provides cushioning and flexibility in several areas of the body. One place it is found is under the skin. Certain organs such as the liver, spleen, and lymph nodes are made of a spongelike kind of connective tissue called reticular tissue. Adipose tissue is connective tissue whose cells have absorbed large amounts of storage fat.

Cartilage is found in several areas of the body. One type is found in the external ear; it is called elastic cartilage. Hyaline cartilage covers the joint surfaces of bones. The discs between the vertebrae of the vertebral column are made up of cushionlike connective tissue called fibrocartilage. Cartilage is sometimes called gristle. It is more dense than most other types of tissue but not as hard as bone. The connective tissue that forms the skeleton is known as bone. Bone may be dense and hard, but it is living tissue. It is made up of bone cells called osteocytes which secrete a strong cementing material called osteoid tissue. Bone contains blood vessels and nerves. Because it is living material, bone is able to repair itself when it breaks. A large part of the material making up the dense part of the bone is composed of calcium and phosphorus as well as other minerals.

Muscle Tissues

The muscular and skeletal systems together provide the major means of support and movement for the body. To most observers the best-known kind of muscle is the skeletal muscle. This is the muscle that controls the movement of the limbs and other parts of the skeleton. It is also partially responsible for the particular shape or form that animals take.

Skeletal muscle is under the voluntary control of the animal. It is able to undergo strong rapid contractions. Skeletal muscle is also referred to as striated muscle due to the appearance of the individual muscle fibers. Each muscle cell, or fiber, has a banded or striped appearance which is caused by the nature of the chemical substance inside the cell. Each fiber of skeletal muscle is long and slender. Many of these cells are needed to form a single muscle. Several muscles are involved in the movement of a single limb. Some cause joints to bend or flex and are called flexors. Other muscles cause the joint to extend and are called extensors.

Another type of muscle is the kind found in the walls of most of the internal organs of the body. It does not have the banded or striated appearance of skeletal muscle, so it is called smooth muscle. It contracts much more slowly and with less strength than skeletal muscle. Most smooth muscle of the body is under the automatic control of the nervous system and endocrine system. For this reason smooth muscle is also called involuntary muscle. The smooth muscle in the walls of the intestines functions to mix the contents for more thorough digestion and to move the contents down the tract at the appropriate time. Smooth muscle in the walls of glands provide them with the ability to eject their secretions.

A third type of muscle is cardiac muscle, which is found in the heart. This muscle has many of the characteristics of skeletal muscle. It is striated, relatively strong, and able to contract quickly. On the other hand, cardiac muscle is under the automatic control of the body, so it is a kind of involuntary muscle. Cardiac muscle begins to function early in the embryonic life of an animal and continues to work until the animal dies.

THE SKELETAL SYSTEM

The skeleton functions in providing strength, support, form, and protection to the body. Skeletal muscles are attached to the skeleton to provide the strength and leverage for running, flying, and other body activities. Many of the internal organs are well protected inside their boney compartments. The form and shape peculiar to specific species is determined to a large extent by the skeleton. For convenience of discussion, the skeleton is divided into three areas: the axial skeleton, the pectoral limb, and the pelvic limb.

The Axial Skeleton

The axial skeleton is made up of the skull, the vertebral column, and the rib cage. It includes all bones except those in the appendages. No attempt will be made here to present the names of all the bones of the axial skeleton. The skull is divided into two areas: the cranium and the face. The cranial bones surround the brain cavity. The bones of the cranium are joined at various joints called sutures.

In those animals that develop horns, the horns grow around a bony extension of the frontal bones. The inner and middle ears are hollowed out of the temporal bones. The bone that forms the floor of the cranium is called the sphenoid bone. The pituitary gland lies in a pocket formed in the sphenoid bone. The occipital bone of the skull is the one that connects to the vertebral column. The spinal cord passes through an opening in the occipital bone called the occipital foramen or foramen magnum. Most of the facial and cranial bones are paired with one on the right and one on the left side of the skull. Neither the sphenoid or the occipital bones are paired.

Several pairs of bones make up the face of an animal, only a few of which will be mentioned. The largest pair of facial bones in most animals are the mandibles, or lower jaw bones. Both the jaw teeth, or molars, and the lower incisors grow out of the mandibles. The upper molars grow out of the maxillae. The upper incisors grow out of the incisive bones. The bridge of the nose is formed by the nasal bones. Some of the bones of the skull have hollow spaces inside them called sinuses. Most of the facial bones as well as the cranial bones are held together by joints called sutures. The joints between the mandibles and the temporal bones of the cranium are freely moving joints. They are both hinge joints and sliding joints. These joints allow the mouth to be opened as when chewing and also permit the lower jaw to slide forward and rearward as when grinding food.

The vertebral column is made up of many bones called vertebrae. The first vertebra is called the atlas. It forms a hinge joint with the occipital bone of the skull. The atlas forms a pivotal joint with the second vertebra called the axis. This pivotal joint allows the neck to be twisted, moving the head from side to side. All of the other vertebrae form slightly moving joints between pairs of vertebrae. The vertebral column is divided into several regions, each region having several vertebrae. The cervical vertebrae are those found in the neck region. There are 7 cervical vertibrae in mammals, and 14 in chickens. A giraffe has no more cervical vertebrae than a dog. The thoracic vertebrae are those that are connected to the ribs. The number of these vertebrae vary from 12 to 18 depending on the species. There is a pair of ribs for each thoracic vertebra. The loin or waist region is called the lumbar area of the vertebral column. The sacral vertebrae are usually fused to form the sacrum. The pelvis is joined to the sacrum. Finally, the last or caudal part of the vertebral column contains the coccygeal vertebrae. Depending upon whether the animal has a tail, the number of these vertebrae will vary from 4 in humans to more than 20 in some tailed species.

The bones of the skull are classified as flat bones, having one short dimension and two longer dimensions. The ribs are also flat bones. The vertebra are irregular-shaped bones.

The Pectoral Limb

Animals with four legs are called quadrupeds. The front pair of legs are called the pectoral limbs. These correspond to the arms of humans. In learning the names of the bones of the pectoral limb, it is interesting to compare those of humans with the domestic animals. Refer to Figure 3-2 for the skeleton of a cow.

The pectoral limb includes the following bones:

Scapula

Humerus

Radius

Ulna

Carpals

Metacarpals

Phalanges

The pectoral limb is connected to the axial skeleton by muscles and connective tissue and not by a bony joint. The joint between the scapula and humerus is a ball-and-socket joint. The movement in cattle, horses, sheep, and hogs, however, is limited by the musculature to not much more than a hinge joint as far as degree of movement is concerned. The joints in the remaining part of the pectoral limb function as hinge joints. The carpus is composed of two rows of small cube-shaped bones that form sliding joints between them, but the overall effect of the entire joint is similar to a hinge joint. The carpus or knee of a quadruped corresponds to the wrist in the human skeleton.

The ulna forms the point of the elbow. It is fused to the radius in the horse. The horse has only one metacarpus, or cannon bone. It corresponds to the third metacarpus in the human hand, the metacarpus of the thumb being number one. Cattle and sheep each have one metacarpus formed by the fusion of the third and fourth metacarpals. The pig has four metacarpals, the first being absent. The number of digits, or toes, varies from one to five in the various mammals. These correspond to the five fingers in the human hand. Horses have one, cows have two, swine have four, and dogs have five. The first digit in cats and dogs is the dewclaw which corresponds to the human thumb. In humans, the clavicle, or collar bone, connects to the scapula. In chickens the clavicle is sometimes called the wishbone.

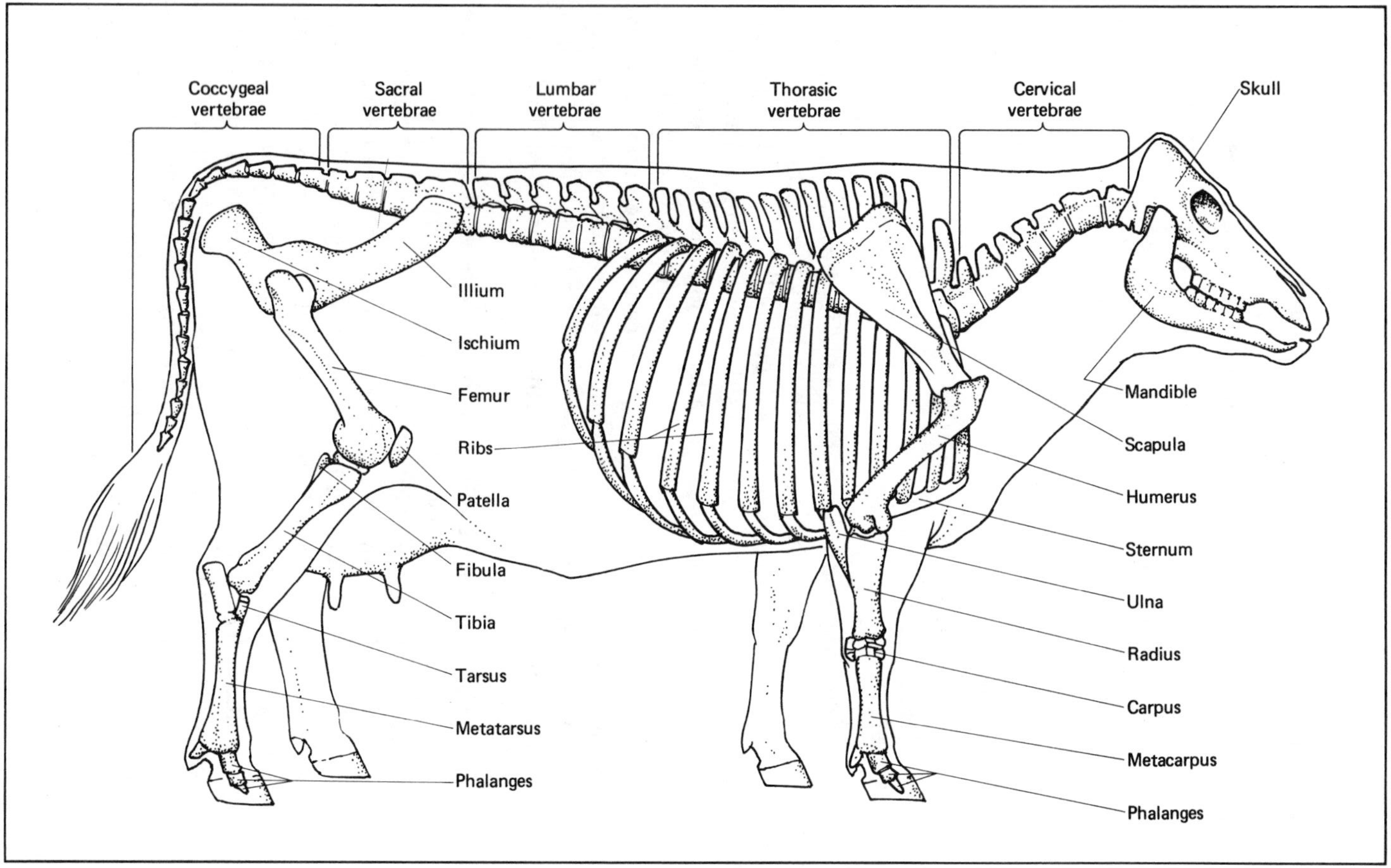

Figure 3-2 Skeleton of a cow.

The Pelvic Limb

The rear legs, or pelvic limbs, are connected to the axial skeleton by means of the pelvic girdle. Three pairs of bones make up the pelvic girdle. They are the illii, ischii, and pubi. Singular forms of these terms are illium, ischium, and pubis, respectively. The pubic bones form the floor of the pelvis. The anterior points of the two illii form the hooks or hook bones in the cow. The posterior ends of the two ischii form the pinbones located on each side of the tail head in the cow. The joint between the pelvis and sacrum is called the sacroilliac.

The remaining bones of the pelvic limb include the following:

Femur

Tibia

Fibula

Tarsals

Metatarsals

Phalanges

The femur connects to the pelvis by means of a ball-and-socket joint. The joint between the femur and the tibia and fibula is called the stifle in four-legged animals; it is the knee in humans. Another bone, the patella, is also present in the stifle joint. In humans it is called the kneecap. The tarsus, or hock joint, is very similar to the carpus in that is consists of two rows of several small bones. The hock corresponds to the ankle in humans. The metatarsus, or rear cannon bone, is very similar to the metacarpus of the front leg. The phalanges of the pelvic limb are essentially the same in structure and appearance as those of the pectoral limb.

Most of the bones of the limbs are classified as long bones. Long bones tend to be somewhat cylindrical with one long dimension. Figure 3-3 shows schematically the structure of a typical long bone. Beneath the epiphysis on each end of the bone in immature animals, there lies a layer of cartilage. It is in these areas where the bone can continue to grow in length. When an animal reaches sexual maturity, the epiphyseal cartilage disappears and the bone can no longer grow

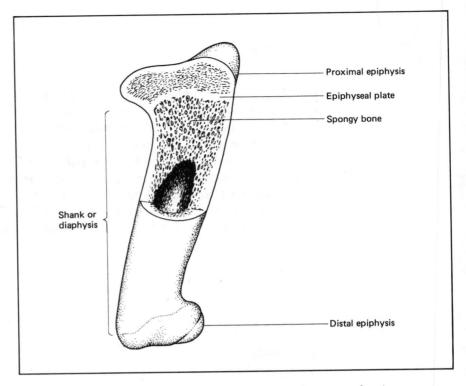

Figure 3-3   Diagram of a long bone, the humerus of a pig.

in length. Dwarfs are sometimes produced with these cartilaginous areas disappear too soon.

INTERNAL SYSTEMS OF THE BODY

All of the major systems within the body will be presented in this section with the exception of the endocrine system, the reproductive systems, and the digestive system. Included are the nervous system, the circulatory system, the respiratory system, and the urinary system. The endocrine system will be presented in the following section. It is the purpose of this book to present only the major features and characteristics of these systems. For more detail, the reader is directed to the many available books on the subjects of anatomy and physiology of both humans and domestic animals.

The Nervous System

The nervous system functions to help the body to be aware of and adjust to various changes in both the internal and external environments. The nerve cells themselves are designed to carry information from one part of the body to another. The nervous system in general is divided into two large regions for purposes of study. The central nervous system is made up of that part located inside the skull and vertebral column. It includes the brain and spinal cord. The peripheral nervous system includes all the nerves that radiate from the central nervous system to all parts of the body.

The central nervous system is the master control center of the whole nervous system. The brain is composed of several parts or regions. The highest level of the brain is the cerebrum. The outer layer of the cerebrum is called the cortex. The highest level and most complex activities and decisions are controlled from this region. The cerebellum is located posterior to the cerebrum (see Fig. 3-4). Compared to the cerebrum and spinal cord, the cerebellum handles activities of intermediate complexity. A major function of the cerebellum is that of coordination of body movements.

The medulla, or brain stem, is the part of the brain that connects to the spinal cord. The medulla also handles activities of intermediate complexity. Some of these activities include the regulation of breathing and heart rate, coughing, sneezing, and control of posture. Nerve fibers that pass up and down the spinal cord and brain stem must pass through the center of the brain called the thalamus. Below the thalamus is a region called the hypothalamus. The pituitary gland is connected to the hypothalamus.

The spinal cord is made up of fibers carrying information from various parts of the body to the brain. It also contains fibers carrying information from the brain to various parts of the body. It can be likened to a major highway carrying traffic in both directions.

The peripheral nervous system is composed of nerves leading to and from the spinal cord and brain. A pair of spinal nerves are connected to the spinal cord at the level of every vertebra. Sensory nerve fibers carry impulses from the sensory organs and receptors into the spinal cord. Depending upon the particular stimulus, the impulses are analyzed and interpreted by the central nervous system. Impulses are carried from the spinal cord over motor fibers to mus-

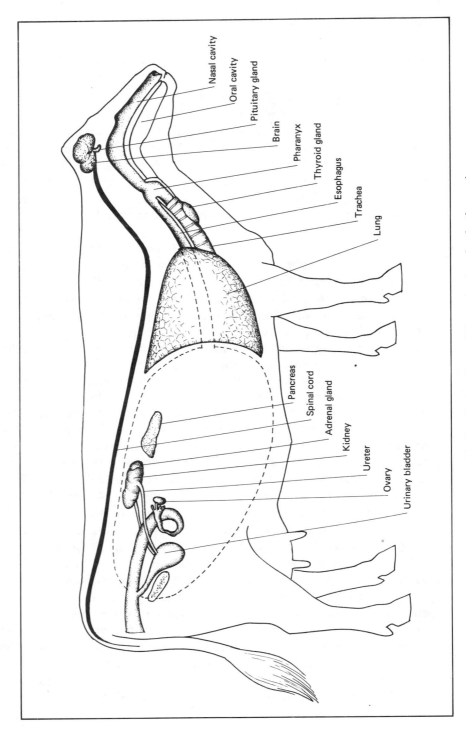

Figure 3-4 Diagram showing the locations of the endocrine glands, the respiratory and urinary organs, and the central nervous system of a cow.

cles or glands to bring about some particular response or adjustment to the environment.

There are twelve pairs of cranial nerves leading to and from the cerebrum and medulla. Some of these nerves carry only sensory fibers such as those leading from the eyes, ears, and nose. Some of the cranial nerves carry only motor fibers such as the ones leading to the muscles that control the movement of the eyes and the tongue. Other cranial nerves carry both sensory and motor fibers, as do each of the spinal nerves.

Many of the activities of the body are under automatic control. The part of the peripheral nervous system that controls such activities is called the autonomic nervous system. For example, when an emergency arises, many of the changes that occur in preparation for coping with the emergency are controlled without any conscious effort on the part of the animal. The heart beats faster, breathing becomes deeper and faster, the eyes are adjusted for distant vision, and blood is diverted from many of the internal organs to the skeletal muscles. These preparations are a part of the "fight-or-flight" reaction of the body. The part of the autonomic nervous system that carries this kind of information is called the sympathetic nervous system. It is also associated with a part of the adrenal glands. As we will see later, the functions of adrenalin are very similar to those of the sympathetic nervous system.

The part of the autonomic nervous system that carries impulses resulting in the relaxation and more restful kinds of responses is called the parasympathetic nervous system. The kinds of activities associated with this part of the nervous system are the opposite of those of the sympathetic system. You would expect to see these reactions when an animal is asleep.

The Circulatory System

The basic parts of the circulatory system are the heart, the arteries, the capillaries, and the veins. The arteries carry blood from the heart to the tissues. Nutrients and waste products are exchanged between the blood and the cells while the blood is in the capillaries. The blood then passes into the veins to be carried back to the heart. The heart is the pumping organ for the circulatory system. It consists of four chambers: two lower chambers called ventricles and two upper chambers called atria. The heart is a very muscular organ, particularly in the region of the ventricles. The heart contracts from the bottom up. The ventricles contract as the atria relax. As the ventricles relax, the atria contract, but with not as much force as the ventricular contractions. See Figure 3-5 for a diagram of the heart.

Blood enters the right atrium of the heart from two major veins. The anterior vena cava carries blood from the head and shoulder regions of the body. The posterior vena cava carries blood from the posterior organs of the body. At the same time that blood enters the right atrium, blood is also entering the left atrium from the pulmonary veins leading from the lungs. Blood enters the two atria as they are relaxed. As the atria contract and the ventricles relax, the blood moves into the ventricles. Contraction of the ventricles then forces blood out of the heart through two major arteries. Blood leaving the left ventricle enters the aorta which ultimately branches into smaller arteries and eventually carries oxygenated blood to all the tissues of the body. Blood from the right ventricle passes into the pulmonary artery and is carried to the lungs to exchange carbon dioxide for oxygen.

The blood serves several functions, some of which can be categorized as protective or maintenance functions. The clotting mechanism of the blood is such a function. Whenever a blood vessel is cut or ruptured, certain reactions occur in the blood that result in the formation of a clot to close the opening and thus prevent the animal from bleeding to death. Buffering agents in the blood keep the pH of the blood from deviating from normal. If the blood becomes too acid or too alkaline, it can upset the normal body chemistry and cause serious problems and even death.

When tissues become low in water, the blood can provide it through the mechanism of osmosis. The blood obtains its fluid supply from the digestive tract. The blood also functions in helping maintain body temperature. The warmer blood moves from the depths of the body organs toward the surface where it is cooled and moves back toward the internal organs. In this regard it is functioning very much like the coolant in an automobile radiator.

The white blood cells, or leukocytes, function in disease prevention and control. More will be said about this particular function of the blood in Chapter 16.

Many of the functions of the blood can be categorized under the heading of transportive functions. Oxygen and part of the carbon

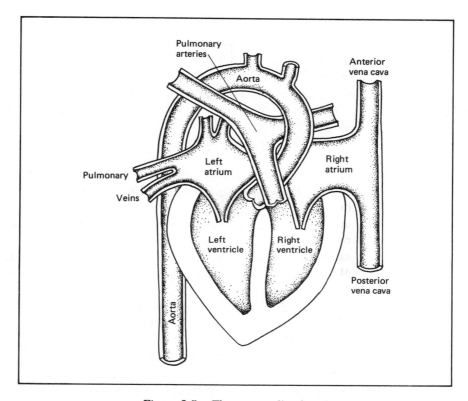

Figure 3-5   The mammalian heart.

dioxide that is produced as a waste product in the body are carried by the hemoglobin of the red blood cells, or erythrocytes. Oxygen forms a loose association with iron in the hemoglobin when the blood passes through the lungs. When the blood passes into the tissues by way of the capillaries, it gives up the oxygen to the cells which require it to carry on metabolism. About one-third of the carbon dioxide is carried on the protein part of the hemoglobin molecules. The balance is dissolved in the plasma of the blood.

Nutrients such as glucose, amino acids, fatty acids, minerals, and vitamins are absorbed from the digestive system and transported to the tissues by the blood. Waste products are carried from the cells and filtered through the kidney and disposed of in the urine. The blood is also the carrier of the hormones produced by the endocrine glands.

The Respiratory System

The two major functions of the respiratory system are to carry oxygen from the atmosphere into the body and to dispose of the carbon dioxide produced by the body. The respiratory system also functions in temperature control. A panting dog expels heat and carries cooler air into the lungs. The air forced from the lungs also is used to produce sounds as it passes through the larynx, or vocal cords.

The respiratory system consists mainly of the nostrils, nasal cavity, pharynx, larynx, trachea, bronchi, and lungs. The oral cavity or mouth may also become involved as a part of the respiratory tract if necessary. The pharynx is the common passageway for the respiratory and digestive tracts. The opening to the larynx is called the glottis. When food passes from the mouth through the pharynx on its way to the esophagus, a valve called the epiglottis closes the glottis to prevent food from passing into the trachea and the lungs.

The trachea branch into two bronchi, each leading to one of the two lungs. The bronchi branch into smaller tubes called bronchioles. The bronchioles branch several times into smaller and smaller bronchioles and finally lead to small sacs called alveoli. The very thin-walled alveoli are surrounded by capillaries which are a part of the pulmonary circulatory system. Air that is inhaled into the lungs contains large amounts of oxygen. The blood in the pulmonary capillaries is low in oxygen. On the other hand, the pulmonary blood is high in carbon dioxide. An exchange of gases occurs between the air in the alveoli of the lungs, and the pulmonary blood. Oxygen enters the blood while carbon dioxide enters the alveoli. When the animal exhales, large amounts of carbon dioxide are expelled.

The Urinary System

The urinary system is made up of those structures that produce, store, and transport urine. In mammals, these structures include the two kidneys, the ureters, the urinary bladder, and the urethra. Water and waste products, as well as many nutrients are filtered by the kidney. Most of the water and nutrients are then reabsorbed and a small amount of water and the waste products are sent to the urinary bladder as urine. The urine is released from the body through the urethra when the bladder becomes full. For the most part, urine is produced continuously by the kidneys and stored in the bladder. In chickens, the ureters lead from the kidneys to the cloaca. The chicken does not possess a urinary bladder or urethra. Urine is voided along with the feces.

The kidneys in most animals are smooth, bean-shaped organs located in the dorsal part of the abdominal cavity. The kidneys in cattle and chickens are lobulated (see Fig. 3-6). The kidneys are well supplied with blood vessels. It is estimated that about 25 percent of the blood passing through the aorta is diverted through the two renal arteries to the kidneys. Most of the blood passing into the kidneys will be filtered to remove waste products. A relatively small part is needed to carry nutrients and oxygen to the kidney tissue itself. The renal arteries branch several times into smaller and smaller arteries that distribute blood throughout the cortex or outer layer of the kidney.

Figure 3-6 shows a diagram of a section of a kidney and includes the filtering mechanism. The blood is carried by very small arteries called arterioles to thousands of small tubules called renal tubules or nephrons. One end of a nephron forms a double-walled, cup-shaped bulb. One of the arterioles enters this bulb, called a Bowman's capsule, where it forms a small tuft of capillaries. Water, nutrients, and waste products are filtered from the blood vessels into the lumen of the nephron. As the filtrate passes down the nephron, most of the water and nutrients are reabsorbed through the walls of the tubule into other capillaries that surround the renal tubule or nephron.

The waste products are not reabsorbed but are carried down the tubule by a small amount of water that is not reabsorbed. The fluid passes from the renal tubule into collecting ducts which lead through the medulla and finally into the pelvis of the kidney. This fluid that enters the collecting ducts from the renal tubules is called urine. The pelvis of the kidney is the common collecting point for all the collecting ducts. The pelvis leads into the ureter which carries the urine to the urinary bladder.

The reabsorption of the renal filtrate is partially under the control of antidiuretic hormone. More will be said about this hormone in the next section.

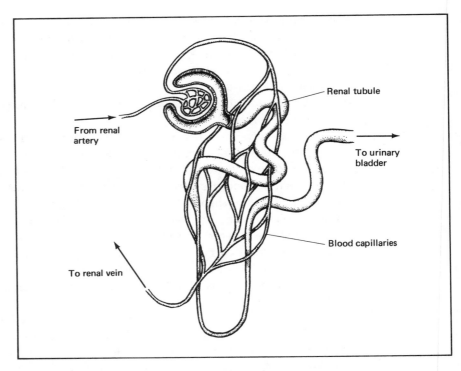

Figure 3-6   Diagram of a section of a kidney showing a renal tubule.

THE ENDOCRINE SYSTEM

An endocrine gland is a gland that secretes a substance that must be absorbed by the blood passing through the gland. It has no duct system to transmit its secretion to some specified area as does an exocrine gland. The secretion of an endocrine gland is called a hormone. Generally, a hormone has a target organ or tissue upon which it has some specific effect. For example, follicle-stimulating hormone (FSH) is carried throughout the body but it has its effect only on the ovaries or testes. On the other hand, growth hormone has an effect on practically all tissues.

Some hormones are called tropic or trophic hormones. This means they stimulate other glands, usually other endocrine glands, to release their hormones or secretions. An example is thyroid-stimulating hormones, or thyrotropic hormone (TSH). It stimulates the release of thyroxine from the thyroid gland. When the level of thyroxine increases in the blood it tends to cause less of the thyrotropic hormone of the pituitary to be released. On the other hand, when the level of thyroxine in the blood is low, it induces the pituitary gland to release more TSH. This is an example of feedback regulation.

The Pituitary Gland

Another name for the pituitary gland is hypophysis. This gland is located in a pocket hollowed out of the floor of the brain cavity. It is structurally a part of the brain. As it can be seen in Figure 3-7, the pituitary gland is connected to the hypothalamus by a stalk giving it the appearance of a small shoe or boot. The hypophysis is divided into two distinct glands called the anterior lobe and the posterior lobe. The anterior pituitary secretes follicle-stimulating hormone (FSH), luteinizing hormone (LH), growth hormone, adrenocorticotropic hormone (ACTH), thyroid-stimulating hormone (TSH), and prolactin. The two hormones released by the posterior lobe are oxytocin and antidiuretic hormone (ADH). In some mammals another hormone, melanocyte-stimulating hormone (MSH), is released by the intermediate lobe which is usually included as a part of the anterior lobe.

Two of the hormones secreted by the anterior lobe are called gonadotropins; they are FSH and LH. Because their function is associated with reproduction, they will be discussed at some length in the chapters on reproduction. Oxytocin is associated with parturition and milk ejection. Prolactin functions in milk secretion. Oxytocin and prolactin will also be discussed later. Both ACTH and TSH are specifically related to the functions of the hormones of their respective target organs, the cortex of the adrenal glands and the thyroid gland.

Growth hormone has the general function of stimulating the growth of all cells. It is particularly effective on bone and muscle tissue. An overproduction of growth hormone tends to cause an increase in the size of growing animals. A deficiency of growth hormone would tend to produce a dwarf. A deficiency

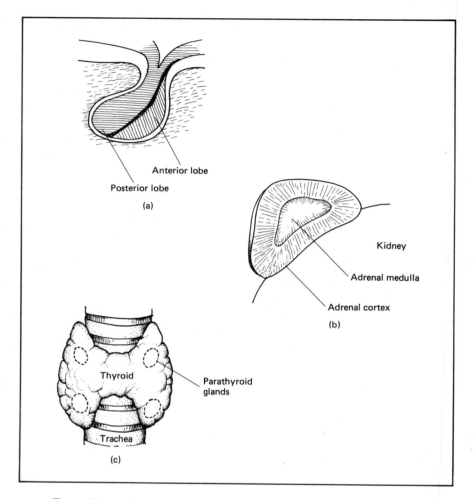

Anterior lobe

Posterior lobe

(a)

Kidney

Adrenal medulla

Adrenal cortex

(b)

Thyroid

Parathyroid glands

Trachea

(c)

Figure 3-7   Diagrams of selected mammalian endocrine organs. (a) Pituitary gland. (b) Adrenal gland. (c) Thyroid and parathyroid glands.

of growth hormone and all anterior pituitary hormones can be brought about by removing the pituitary. This operation is called a hypophysectomy.

Antidiuretic hormone stimulates the increased permeability of the renal tubules, or nephrons, of the kidneys. After the fluid has been filtered into the tubules, recall that much of the fluid and the dissolved nutrients are reabsorbed before urine is formed. The effect of ADH is to increase the amount of filtrate that will be absorbed, thus reducing the amount of urine produced. The term diuresis refers to urine production. Antidiuresis refers to the inhibition of urine production.

Antidiuretic hormone has the overall effect of controlling the amount of water in the blood. A decrease in the water level of the blood causes an increase in the concentration of substances dissolved in the blood. This changes the osmotic pressure of the blood. The change in osmotic pressure results in the release of ADH from the posterior lobe of the pituitary. The ADH stimulates an increased reabsorption of water from the renal tubules resulting in the restoration of the water level of the blood.

Antidiuretic hormone has a second function in that it causes an increase in blood pressure. Another name for ADH is vasopressin, referring to this second function.

The Thyroid and Parathyroid Glands

There is apparently no functional relationship between the thyroid and parathyroid glands. In most mammals, however, they are associated structurally. As shown in Figure 3-7, the thyroid gland is located in the neck region near the larynx and trachea. There are usually two lobes, one located on either side of the larynx. The parathyroid glands are usually embedded in the thyroid tissue or located very near the thyroid gland. There are usually four or more separate nodules, or actually four separate parathyroid glands.

The usual hormone of the thyroid gland is called thyroxine. It stimulates cells to work faster by increasing their rate of metabolism. Thyroxine generally affects all cells in the body. In the synthesis of thyroxine by the thyroid gland, the mineral iodine is required. Without iodine, the substance produced has no thyroxinelike activity. In situations where iodine is deficient, the thyroid gland enlarges in a futile attempt to produce more of the hormone. This abnormal enlargement is called a goiter.

Increased production of thyroxine is called hyperthyroidism. An animal with this condition will show nervousness, increased appetite, and a general loss in weight due to the increase in metabolic rate. Hypothyroidism results in a decrease in thyroxine production causing a decrease in metabolism, a tendency to gain weight, and a reduction in general body activity. Lethargy is the term used to describe the slow, lazy-seeming activity of hypothyroidism.

The release of thyroxine is under the control of thyroid-stimulating hormone (TSH). Recall that a feedback regulatory mechanism exists between these two hormones. A decrease in the level of thyroxine in the blood results in the release of TSH. The TSH in turn stimulates the release of thyroxine. The increased level of thyroxine tends to inhibit the release of more TSH from the pituitary.

Parathyroid hormone, or PTH, is also called parathormone. It functions in maintaining the calcium level of the blood. When the calcium level drops below normal, it causes the parathyroid glands to release PTH. The PTH in turn stimulates the uptake of calcium from the digestive system and the release of calcium from the bones into the blood in an attempt to restore the blood calcium level.

A deficiency of PTH production or the removal of the parathyroid glands results in a low blood level of calcium, called hypocalcemia. Low blood calcium affects the sensitivity of nerves and muscles. If the deficiency is severe enough, it can lead to tumors and twitching of the muscles and eventually to tetany, or sustained muscle contraction, and death. An overproduction of PTH causes an increase in blood calcium, or hypercalcemia, and a corresponding weakening of the bones due to a loss of calcium.

The Adrenal Glands

The term adrenal indicates a relationship to the kidneys. The adrenal glands in most mammals are located very close to, or may actually touch, the kidneys. Each gland is divided structurally into two parts, the adrenal cortex and the adrenal medulla. The medulla produces the two hormones called adrenalin and noradrenalin. The cortex produces several hormones, which are categorized into two groups: the glucocorticoids and the mineralocorticoids.

The hormones adrenalin and noradrenalin have very similar functions, although they are not identical. Another term for adrenalin is epinephrine. Noradrenalin is also called norepinephrine. The functions of adrenalin are related to the "fight-or-flight" reactions associated with the sympathetic nervous system explained earlier. In fact, the sympathetic nervous system is connected to the medulla of the adrenal glands. This is a good example of an interrelationship between the nervous system and endocrine system.

The glucocorticoids produced by the adrenal cortex are under the control of ACTH from the anterior pituitary in much the same way that thyroxine is controlled by TSH. The glucocorticoids, as the term suggests, are involved in making glucose available for cell metabolism. They also tend to make the body less sensitive to certain stresses. Glucocorticoids include cortisol, cortisone, and corticosterone.

The mineralocorticoids are involved in maintaining the level of water and certain minerals in the blood. One of the most effective hormones of this kind is called aldosterone. The pituitary hormone ACTH does not seem to have as much effect upon the release of the mineralocorticoids as on the glucocorticoids.

The Pancreas

The pancreas is located next to the small intestine in the region of the duodenum. One of its functions is to secrete digestive enzymes which are carried into the small intestine by a tube or duct. In this function it acts as an exocrine gland. Located throughout the tissue of the pancreas are clumps or clusters of cells called the islets of Langerhans. These are the hormone-producing cells of the pancreas. The two hormones produced here are insulin and glucagon.

Insulin is the more important of the two hormones of the pancreas. It functions in maintaining the glucose level of the blood. When glucose levels rise too high the pancreas is stimulated to release insulin. The insulin then stimulates the body tissues to absorb more glucose which results in the lowering of the sugar level of the blood. As the blood sugar level returns to within the normal range it tends to inhibit the release of additional insulin. The mechanism where blood sugar levels control the release of insulin from the pancreas is another example of feedback regulation. A deficiency of insulin will

result in an increase in blood sugar level. This is called hyperglycemia. Injections of additional insulin result in the opposite effect,
hypoglycemia.

The actual role of the hormone glucagon is not clear, although
it is known to cause an increase in blood sugar level when released.
In this sense, it can be thought of as antagonistic to insulin.

STUDY QUESTIONS AND EXERCISES

1. Define each of the following terms:

ACTH	cytoplasm
adrenalin	dermis
alveolus	diaphragm
anterior	distal
antidiuresis	diuresis
anatomy	dorsal
aorta	embryology
appendage	endocrine
ADH	epinephrine
artery	epiphysis
atlas	epithelium
atrium	erythrocyte
axis	femur
axon	foramen
cardiac	gland
caudal	glucagon
cartilage	goiter
cell	gonadotropin
cerebellum	gristle
cerebrum	hemoglobin
cervical	histology
collagen	hormone
cortex	humerus
cranial	hyperglycemia
cytology	hypocalcemia

hypodermis   pharynx
hypoglycemia   physiology
hypophysectomy   plasma
hypophysis   posterior
hypothalamus   prolactin
insulin   proximal
islets of Langerhans   PTH
larynx   quadruped
lateral   reflex
leukocyte   sagittal
lumbar   sensory nerve
lymph   tendon
mandible   tetany
maxilla   thalamus
medial   thoracic
medial plane   thyroid
medulla   thyroxine
melanin   tissue
motor fiber   trachea
mucus   transverse
nephron   TSH
neuron   urine
organ   vascular
osteocyte   vasopressin
oxytocin   vein
pancreas   vena cava
parathormone   ventral
pH   ventricle

2. Indicate the scientific name for the study of each major system of the body and the major organs involved in each.

3. What would you study in a course dealing with the comparative systematic anatomy and physiology of domestic animals?

4. Name the various body cavities and the organs located in each.

5. How many kinds of tissues are present in the body? Name them.

6. Name some of the materials found in the skin.

7. List some examples of modified epithelium.

8. List and compare several kinds of connective tissue.

9. Name several bones of the cranium and face of the cow and indicate a feature of each that should help identify it.

10. List six classifications of joints of the skeleton.

11. Name the five regions of the vertebral column.

12. List the bones and joints of the pectoral limb of the horse.

13. List the bones and joints of the pelvic limb of the pig.

14. Compare the number of metatarsals and phalanges in the various domestic mammals and humans.

15. Describe and draw the structure of a typical long bone.

16. What are the functions of the muscular system?

17. Compare the three types of muscle tissue.

18. Compare the functions of motor and sensory nerve fibers.

19. Compare the functions of the sympathetic and parasympathetic nervous systems.

20. List some of the maintenance functions of the blood.

21. List some of the transportive functions of the blood.

22. Compare the nature and function of circulation associated with the left and right sides of the heart.

23. Describe and diagram the structure of the respiratory duct system.

24. Describe the functions of the respiratory system.

25. Describe and diagram the structures of the urinary system.

26. Explain the functioning of a nephron.

27. Compare the structure of endocrine and exocrine glands.

28. What is a tropic hormone?

29. Explain the negative feedback mechanism in endocrinology.

30. Describe the parts of the pituitary gland and list the hormones released by each part.

31. List the general functions of each of the pituitary hormones.

32. Describe the structural relationship between the thyroid and parathyroid glands.

33. What is the function of thyroxine and what it its relationship to iodine?

34. Explain the relationship between the pituitary gland and the thyroid gland.

35. What determines the secretion level of parathyroid hormone?

36. List the hormones produced by the adrenal glands and indicate their functions.

37. Describe the structure and overall functions of the pancreas.

4

Animal Behavior

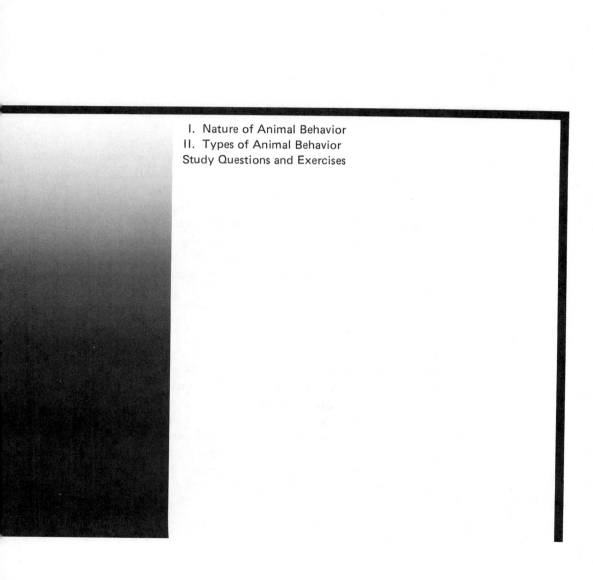

Animal behavior is defined as the response of an animal to changes in the environment. It is a complex reaction that involves interactions between the animal's genetic makeup and its many experiences. Knowing how an animal will react under a given set of circumstances could be helpful in trying to change certain types of behavior to the advantage of the producer. Observing abnormal behavior in an animal can be suggestive of a particular disease problem or poor management practice.

NATURE OF ANIMAL BEHAVIOR

A part of an animal's behavior pattern is its ability to communicate with others of its species and, in some cases, with humans. One important way that animals communicate is with the use of sound. The ability to hear is well developed in most species of domestic animals. Distress calls, the notifying of other animals that food has been found, courting sounds, and chatter between mother and young are several examples of sound communication among animals. Cattle, hogs, horses, and dogs may make sounds to their masters or herders when hungry or distressed. Visual methods of communication are also common among domestic animals. Dogs and cats will arch their backs and cause the hair to stand up when they become involved in hostile situations. Male birds will strut and waltz in courtship displays for the females of the species. The sense of smell is better developed in most domestic animal species than it is in humans. Male mammals can detect females in estrus by utilizing their senses of smell. Dogs and stallions may use urine or feces as scent markers to identify their territory.

One of the causes for the way an animal behaves is genetic. The behavior patterns that an animal inherits are said to be instinctive in nature. Newborn mammals, for example, do not have to learn the suckling behavior. To the extent that behavior patterns are inherited, they can be modified by genetic selection. Many activities of animals may be controlled to some extent by the genes, but must be developed by learning or training. A Thoroughbred foal may have the necessary genetic makeup to someday win the Kentucky Derby, but will never be able to accomplish that task without the proper training. On the other hand, the genetic makeup can be the limiting factor. The best possible training available would never prepare a Belgian stallion to win the Kentucky Derby. So, genetic factors are certainly very important in determining behavior patterns, but environmental factors and interactions between heredity and environment are also important.

Much of how an animal behaves is brought about by what the animal learns from various experiences. The simplest type of learned behavior is called habituation, learning to respond without thinking. The next level of learned behavior is called conditioning. Milk cows become so conditioned to being milked after entering the milking parlor that after a period of time they will begin to "let down" their milk in response to this stimulus. Conditioned behavior can be reinforced by rewarding an animal after it has made a successful response by giving it an apple or a sugar cube, for example, or even just a gentle pat on the neck. Conditioned behavior can also be negatively reinforced by the use of punishment following an unsuccessful response.

One of the highest forms of learning among domestic animals is insight learning, the ability to make the correct response to a stimulus the first time it is encountered. A similar kind of behavior in humans is called reasoning. It is similar to trial and error, except that with trial and error the animal may make several unsuccessful attempts before performing the correct one. An example of insight learning is a chimpanzee obtaining a banana that is out of reach by stacking boxes beneath it and climbing up. Another complex form of learned behavior is called imprinting, a form of social learning. When a gosling or duckling is exposed to any moving object within 36 hours after hatching, it will adopt that object as its parent. The object usually is its mother, but it may be a dog, a cat, a bird of another species, a human, or even some nonliving moving object.

TYPES OF BEHAVIOR

Farm animals exhibit several types of behavior including ingestive, eliminative, sexual, mother-young, combative, gregarious, investigative, and shelter-seeking, behaviors.

1. **Ingestive Behavior.** Eating and drinking are ingestive behaviors that are common to all species of all ages. The first behavior of this type observed in young mammals is suckling. Other examples of ingestive behavior include the grazing habits of ruminants, rooting by pigs, and the pecking habits of birds. Cattle spend about one-third of their time grazing and almost that much time ruminating, or chewing their cud. Dogs and cats drink by lapping fluids with their tongues while cattle and horses make use of a sucking action.

2. **Eliminative Behavior.** As animals vary widely in their ingestive techniques, they also exhibit varying ways of eliminating feces and urine. Cattle, sheep, and chickens tend to deposit feces at random and wherever they happen to be when the need to do so arises. Most other domestic animals tend to defecate in certain specific locations. Pigs are very clean in their defecating habits in that they tend to keep their bedding dry and clean. Cats bury their feces and urine. Male dogs have the habit of lifting one hind leg and urinating on some upright target.

3. **Sexual Behavior.** The sexual behavior of animals has been studied about as much or more than any other form of behavior, including ingestive behavior. Both of these forms of behavior have

very important applications from a commercial and economic standpoint. Sexual behavior involves courtship and mating. Since reproduction is one of the most important functions associated with livestock production, it is important that the livestock producer have an excellent understanding of sexual behavior. Male mammals generally smell and lick the external genitalia of the females in estrus. During estrus, the female will generally accept the advances of the male. Females that are not in estrus will usually not respond to such advances even if they are made by the male. Sexual advances are seldom ever made by males toward females that are not in estrus, at least not in the case of nonprimates. The cock, or rooster, will usually perform a special waltz or dance before attempting to mate. The hen may ignore the cock, she may run away, or she may squat and allow him to mount (see Fig. 4-1).

4. **Mother-young Behavior.** The behavior patterns that occur between mothers and their young begin before the young are actually born and continue beyond weaning. Sows have a tendency to build a nest a few days before the time for parturition to occur. A cow becomes nervous and leaves the rest of the herd for a secluded spot in which to have her calf. After the calf is born, the cow licks the calf and moos gently as if she were talking to it. Mares and ewes act toward their young in a similar fashion. Sows do not attempt to clean their pigs after birth. Mothers of all species are quite protective of their young. If a calf, lamb, foal, or piglet is threatened, the mother will usually attempt to protect it even if it appears to be a futile effort.

One of the first things a newborn mammal attempts to do after birth is to nurse (Figs. 4-2 and 4-3). A newborn lamb, foal, or calf is very unsteady on its feet for the first few hours, but becomes steadily stronger after nursing a few times. After several hours, a cow may leave her calf and return to the herd to graze. The calf will lie very quietly until the mother returns. After a few days, the cow will usually take her calf with her into the herd. The mother is able to recognize her baby by smell, sight, and sound, although this ability does not appear to be as strong in sows as it is in cows and mares. Weaning time is often a traumatic experience for both mother and young. Beef calves are usually weaned at about seven months of age. When they are separated from their mothers they will bawl for several days. Their mothers will usually bawl

Figure 4-1   The waltz or dance performed by a cock toward a hen, an illustration of sexual behavior. [James V. Craig, *Domestic Animal Behavior.* (Englewood Cliffs, N.J.: Prentice-Hall, Inc., 1981) p. 59]

for several days, also. Dairy calves normally do not go through this experience since they are generally removed from the cow within a few hours after they are born.

Mother hens exhibit most of the instincts of care and protection toward their chicks that mammalian mothers show toward their young offspring. These instincts are not observed so much in commercial poultry operations anymore due to the fact that mechanical incubators have replaced the setting hen. Broodiness is the tendency of hens to sit on a nest of eggs. Mother hens will protect their chicks and keep them warm by covering them with their wings. When the hen finds food, she will cluck to her chicks and they will come running at her call.

Figure 4-2   Nursing behavior in pigs. [James V. Craig, *Domestic Animal Behavior*. (Englewood Cliffs, N.J.: Prentice-Hall, Inc., 1981) p. 343]

Figure 4-3   A calf nursing its mother, an illustration of mother-young behavior. (Beefmaster Breeders Universal)

5. Combative Behavior. Also called agonistic behavior, combative behavior includes fighting and other related behavior involving conflict. Males are much more likely than females to take part in combative behavior. Animals that are raised together are less likely to fight among themselves than are strange animals when

they are put together (Fig. 4-4). Fighting will usually occur among a group of strange bulls until a particular "peck" order has been established (Fig. 4-5). Among a group of four bulls, for instance, one bull will be able to dominate the other three. A second bull will dominate two of the others, but be subordinant to the first. The third bull will be subordinant to the first two but will dominate the fourth. The fourth bull is dominated by the other three. When several strange roosters or hens are put together in a pen, the same kind of pecking activity occurs. In fact, the term "pecking order" came about from observations of combative behavior among birds. Cows, sows, and mares will also establish a peck order within individual groups, but their fighting is not nearly as intense as that observed among the males.

6. Gregarious Behavior. Gregarious behavior relates to the social attachments that certain species have one for the other within the species. It is these social attachments that result in animals coming together in herds, flocks, and droves. Sheep show this tendency probably as strongly as do any of the other domestic animal

Figure 4-4   Fighting among weanling pigs that were put together as strangers at four weeks of age. [James V. Craig, *Domestic Animal Behavior*. (Englewood Cliffs, N.J.: Prentice-Hall, Inc., 1981) p. 180]

Figure 4-5 Combative behavior between two Hereford bulls. (Webbphotos, St. Paul, Minnesota)

species. Domestication itself as well as many of the common practices of modern animal husbandry tend to interfere with the natural flocking instinct to some extent. Many animals have the tendency to follow the lead of another animal; for example, if one sheep breaks away from a flock and runs, the rest of the flock generally will follow (see Fig. 4-6). Pigs, dogs, cats, and chickens tend to huddle together to keep warm when the weather is cold.

Figure 4-6 The flocking tendency of sheep is an example of gregarious behavior. (University of Missouri)

Figure 4-7 Investigative behavior among Santa Gertrudis calves. (Missouri Beef Cattleman)

7. **Investigative Behavior.** Investigative behavior involves curiosity on the part of the animal, and a tendency to explore the environment, especially when the environment is new and strange to them. Pigs, turkeys, horses, and cattle seem to be more curious than timid about investigating strange objects than are sheep and chickens. The younger animals of any species are generally more curious than their parents. Figure 4-7 illustrates curiosity on the part of some young bulls toward the photographer. This tendency toward overinquisitiveness on the part of the young animals often causes much worry to the watchful mothers. If there is an opening in a fence or corral, or if a gate has been left open, anyone who raises cattle knows that it will very quickly be found by some curious cows or calves.

8. **Shelter-seeking Behavior.** The tendency of animals to protect themselves from predators, insects, and the weather leads to shelter-seeking behavior. Even though cattle are not as sensitive to severe weather as are some other species of domestic animals, they do attempt to seek protection in times of bad weather. This protection may take the form of timber, brush, valleys, the downwind side of a hill, or any other kind of a windbreak. During periods of blowing rain or snow, cattle head away from the direction of the wind. Sheep also head away from a storm and tend to bunch closely together for protection. Horses possess this same instinct of facing away from the wind. Hogs are particularly sensitive to extremes of both cold and heat (see Fig. 4-8). During hot weather, almost all animals have the tendency to seek shade and water.

112

Figure 4-8   Hogs wallowing in the mud in an attempt to keep cool on a hot day. [James V. Craig, *Domestic Animal Behavior.* (Englewood Cliffs, N.J.: Prentice-Hall, Inc., 1981), p. 237]

STUDY QUESTIONS AND EXERCISES

1. Define each of the following terms:

   agonistic   habituation
   broodiness   instinct
   conditioning   imprinting
   gregarious

2. Why is it important to understand some of the basic principles of animal behavior?

3. How do animals communicate with each other?

4. What are some of the causes of animal behavior?

5. How can behavior patterns be inforced, or reinforced?

6. Compare ingestive behavior among the domestic animal species.

7. Compare eliminative behavior patterns among domestic animal species.

8. How is sexual behavior related to the economic importance of animals?

9. Give some examples of mother-young behavior.

10. Explain what is meant by "peck" order.

11. Describe investigative behavior in domestic animals.

12. How do cattle attempt to protect themselves from bad weather?

5

Principles
of Animal Genetics

Genetics is that branch of biology that deals with genes and their transmission from one generation to the next and their effects on external traits and characteristics. It deals with heredity and variation. As with other subject matter areas, genetics has a language of its own.

Many kinds of traits in animals are controlled by genes. The amount or degree of control varies from trait to trait and from species to species. Environmental factors also play a role in determining how genes express themselves. The many colors and color patterns in animals are inherited characteristics. Color inheritance is used extensively in the explanation of how genes interact and why certain genetic ratios are important. However, we will find that genetics is much more than dominance and recessiveness and 3 : 1 ratios. Knowing how to apply genetic principles helps us to eliminate genetic defects, to select for resistance to diseases and stress, and to increase the productive ability of domestive animals.

This chapter deals primarily with the chemistry, biology, and mathematical bases of inheritance. It deals further with how genes are distributed from one generation to the next, and how they function in one- and two-pair crosses.

GENES AND CHROMOSOMES

In the nondividing cell, chromosomes are relatively long, slender, threadlike structures found inside the nucleus of the cell. They cannot be seen as separate structures with the aid of a microscope until the cell is in the preparatory stages of division. Chromosomes are important to the study of genetics because they contain the genes. The number of genes that can be found on a single chromosome is not known definitely but estimates vary from less than 100 to more than 1000.

Chromosome Number

The number of chromosomes in animal cells varies from species to species. Within a species, however, the number remains constant. In fact, if for some reason the number of chromosomes should vary from normal, vital functions are usually affected to the extent that the organism generally dies during embryonic development. The typical number of chromosomes in the body cells is indicated for the following common animals:

Pig	38
Cat	38
Rat	42
Rabbit	44
Human	46
Sheep	54
Cattle	60
Goat	60
Donkey	62
Horse	64
Dog	78
Chicken	78

The size of chromosomes also varies. They may be very short with only a few genes or be relatively long with a large number of genes. Another distinguishing feature is the centromere which serves as a point of attachment for the spindle fiber during cell division. The location of the centromere also varies among chromosomes. It may be somewhat centrally located or be located nearer to one end than the other. All of these differences in the morphology of the chromosomes make it possible to distinguish one from another when they are observed in a dividing cell. A diagram of a typical chromosome is shown in Figure 5-1.

In addition to the fact that there is a constant number of chromosomes in the body cells of animals, the chromosomes exist in pairs. The members of each of these pairs are the same shape and length and have their centromeres located in the same position. It is often customary to express chromosome number in terms of the number of pairs typical for any species, such as 30 pairs in cattle, 19 pairs in swine, 39 pairs in dogs, or 23 pairs in humans. The two members of the paired set of chromosomes are called homologs of each other. Homologous chromosomes are described as two chromosomes that are alike in size, shape, and position of their centromeres. Cells that contain chromosomes in pairs are said to be diploid.

Chemical Nature of Genes and Chromosomes

Chromosomes are made up of a complex combination of protein and deoxyribonucleic acid (DNA). The protein acts as a matrix or skeleton to support the DNA. The DNA consists of many molecules connected in chainlike fashion along the length of the chromosome. Each gene is thought to be composed of one molecule of DNA. Therefore, if a particular chromosome contains 200 genes, it would in reality contain 200 molecules of DNA.

Genes carry out their functions by controlling the synthesis of substances called enzymes. Enzymes are made of protein and usually function as catalysts in chemical reactions. The activities of a cell are controlled by many hundreds of chemical reactions. The activities of all cells working together in tissues, organs, and systems ultimately cause the animal to have a certain appearance or to behave in a certain manner. This appearance or manner of behavior is called the animal's phenotype.

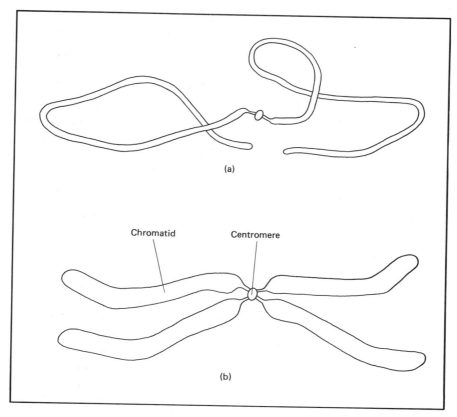

(a)

Chromatid   Centromere

(b)

Figure 5-1 Schematic diagrams of a chromosome before and after duplication. (a) A single chromosome as it might appear in an interphase cell except that it probably would be somewhat longer and thinner than shown here. (b) The same chromosome as above during the prophase and metaphase stages of cell division. The chromosome strands duplicate and become much shorter and thicker than during interphase.

Alleles and Loci

The exact position or location of any gene on a particular chromosome is called a locus. Every gene has a locus. For every locus on a particular chromosome there is an identical locus at the exact position on the homolog. This means that the gene that occupies a locus on one chromosome could also occupy the same locus on the other homologous chromosome. However, the genes that occupy these two homologous loci (plural for locus) may or may not be the same. Whether they are the same or different, they will affect the same set of traits. Further, the genes that occupy these homologous loci will not be found on any of the hundreds of other loci on all the other chromosomes in the cell. If the genes at these homologous locations are different, they will have a different effect upon the set of traits that they control. The different genes that can occupy the same loci on a pair of chromosomes are called alleles or allelomorphs. Keep in mind the key parts to the definition of alleles: The genes must be on the *same locus* of two homologs, they must be *different*, and they affect the *same trait*. For example, one gene may cause a cow to possess horns, and the allele of that gene may cause it to be polled (without horns). The presence or absence of horns is the trait. Two phenotypes may exist for this trait: the horned phenotype, or the polled phenotype. Genotype refers to the genes an animal possesses, while phenotype refers to the way the genes cause the animal to appear or behave.

When describing genes and solving genetic problems, it is necessary to use symbols to represent the genes. The most common way is to use letters for gene symbols. There is no standard or accepted pattern to follow, but one way is to use upper and lower case letters to represent alleles if only two alleles are involved. Suppose at a given locus, the possibility of two alleles exists. We may call this the *A*-locus and use the letters *A* and *a* to symbolize the two alleles. Either allele can occupy either locus, so in a group of several animals we might well expect to see the two alleles present in all possible combinations. If an animal possesses two *A*-genes, its genotype would be written as *AA*. If another animal possesses two *a*-genes, its genotype would be written as *aa*. In these two cases, the genotype of each animal is made up of two identical genes. These genotypes are said to be *homozygous*. A third possible genotype may exist. It will be made up of one of each of the two alleles (*Aa*). This genotype is said to be *heterozygous*.

Before we consider how genes are transmitted from one generation to the next, the subject of cell division should be discussed. An understanding of how cells divide, and how chromosomes duplicate and are transmitted to daughter cells is vital to a fuller understanding of gene segregation and recombination.

Two types of cell division occur in animals. One type involves a single division of a diploid cell to produce two identical diploid daughter cells. This kind of cell division is called mitosis. An understanding of mitosis will help in explaining the second kind of cell division called meiosis. Meiosis involves a series of two divisions whereby four daughter cells are produced from a single diploid parent cell. The four meiotic products are not identical and contain only one-half the number of chromosomes found in the parent cell. The daughter cells are said to be haploid.

Mitosis

The simpler of the two kinds of cell division, mitosis, occurs among the body cells. Body cells, which are also called somatic cells, include all cells except the sex cells or gametes. Mitosis occurs in order to increase the number of cells in growing animals or to maintain the number of cells in tissues as old cells wear out and die. It is not known what causes a cell to begin the process of division. The non-dividing cell is usually described as being in interphase. This phase in the life of a cell ends when preparations for mitosis begin. For ease of discussion the process of mitosis is divided into four phases: prophase, metaphase, anaphase, and telophase.

1. **Prophase** includes most of the activities that prepare the cell for division. In this phase, each chromosome manufactures a complete new chromosome strand. The two strands are called chromatids, and are united as one structure by a common centromere. The DNA, and thus the genes, in each chromatid are identical to one another. Figure 5-1 shows how a chromosome might appear before and after the duplication process. From a genetic standpoint, chromosome duplication could very well be considered the most important aspect of cell division.

The other activities that are included as a part of prophase occur mainly to bring about an orderly division of the doubled chromosomes and ultimately of the cell itself. The chromosomes become shorter and thicker than they were in the interphase cell. The centriole, which is located just outside the nuclear membrane, divides and begins to migrate to two sides of the cell. As the centrioles move, the nuclear membrane disappears. By the time the centrioles have completed their migration to what is referred to as the poles of the cell, the spindle has also formed. The spindle is made up of several fibers that seem to extend from centriole to centriole. Each spindle fiber also appears to connect to a centromere of a doubled chromosome. These stages of prophase along with the other phases of mitosis are illustrated in Figure 5-2.

2. **Metaphase** is also a predivision phase of mitosis. In this phase, the doubled chromosomes become oriented along the equatorial plane of the cell. With the centrioles representing the poles of the cell, an imaginary line running through the cell from pole to pole would represent the axis. The equatorial plane of the cell would be a plane perpendicular to the axis of the cell. At this point in mitosis, the chromosomes are ready to divide.

3. **Anaphase** is the name given to the stage of mitosis where the centromeres divide longitudinally between the chromatids of their respective chromosomes and the separated chromatids move to either pole of the cell. The two chromatids from each chromosome appear to be drawn to opposite poles of the cell by the spindle fiber that connects its centromere to the centriole. Each chromatid can now be called a chromosome.

4. **Telophase** is the final stage in the mitotic process. After the newly formed chromosomes have completed their migration, forming a cluster next to each centriole, the cell membrane constricts between them and eventually separates completely, forming two cells. As this construction and division of the cell procedes, a nuclear membrane forms around the cluster of chromosomes in each half of the cell. The chromosomes again become long and indistinguishable. The two new daughter cells return to the interphase, or nondividing stage. The new cells are both diploid and are genetically identical to each other and to the parent cell from which they were formed. The production of two diploid cells from one diploid cell is made possible by the fact that the chromosomes duplicate prior to the actual division of the cell.

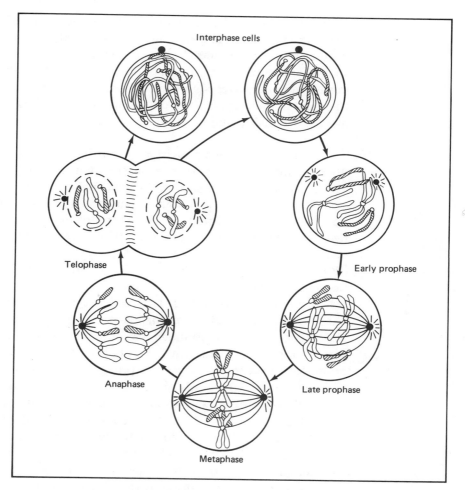

Figure 5-2   A cell undergoing mitosis.

Meiosis

The second type of cell division occurs with certain specialized cells located in the testes and ovaries. The process begins with diploid cells called gonial (germ) cells, ovigonia in the ovaries and spermatogonia in the testes. Meiosis occurs as a part of the processes of gamete formation in males and females. Ovigenesis in the female is explained in Chapter 8; spermatogenesis in the male is explained in Chapter 9.

In this chapter we will be concerned primarily with what happens to the chromosomes. The overall process involves two divisions and results in four products from one original gonial or germ cell. The four meiotic products are haploid—they possess only one-half the normal number of chromosomes present in the normal body cells of the animal. Meiosis occurs only in connection with the production of male and female gametes.

1. **The first meiotic division** begins in much the same way as it does in mitosis. The activities described in the prophase of mitosis also occur in the prophase of the first meiotic division (prophase I). Again, probably the most significant activity from a genetic standpoint is the duplication of the chromosomes. The chromosomes become short and thick; the centriole divides and the two new centrioles migrate to the poles of the cell; the nuclear membrane disappears; and a spindle fiber forms. In addition, the doubled homologous chromosomes seek out one another and appear to lie side by side in the cell. This is called synapsis. The two doubled chromosomes of each homologous pair lying side by side give the appearance of a four-stranded structure called a tetrad. The tetrad gets its appearance and its name from the four chromatids of the two homologs. In a meiotic cell in prophase I in swine, for example, there would be 19 tetrads since the cells of swine contain 19 pairs or 38 total chromosomes. The process of synapsis is sometimes called tetrad formation. Meiosis is illustrated diagramatically in Figure 5-3.

Before the first division actually occurs, the tetrads become oriented along the equatorial plane of the cell in a manner similar to the metaphase of mitosis. In meiosis, this is called metaphase I.

During the period prior to the first division which the homologous chromosomes are in synapsis, a phenomenon called crossing-over occurs. A chromatid of one homolog will lie across one of the chromatids of the other homolog, forming an X-shaped appearance that can be observed under the microscope. This crossover usually occurs at the same locus on each chromatid. When the homologous chromosomes separate during anaphase I, they very often exchange parts of their chromatids. Genes that were once on one chromosome can be found on the homologous partner. This is not an unusual phenomenon in meiosis—it is a natural part of the process. Crossing-over plays an important role in the "mixing" of the genes on the homologs and in the random distribution of genes in the gametes.

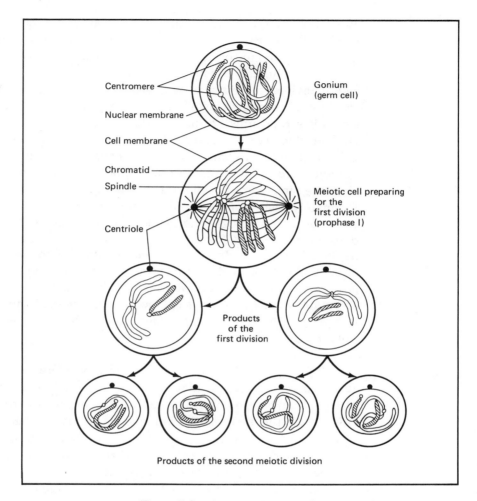

Figure 5-3 A cell undergoing meiosis.

The events of anaphase I of meiosis are somewhat different from those of mitotic anaphase. Recall that in mitosis there was no tetrad—the homologous chromosomes were distributed at random within the cell. During the anaphase the centromeres of each chromosome divided and separated resulting in the formation of two new chromosomes from the two chromatids of each doubled chromosome. In anaphase I of meiosis, however, this does not occur. Instead, the homologous chromosomes become "unsynapsed"—they separate and move toward the poles of the meiotic cell. Each chromosome still possesses two chromatids. As the cell continues to divide, a new nuclear membrane forms around each of the two clusters of chromosomes and the cell constricts between them to form two new cells. This phase of the division is called telophase I. The completion of telophase I produces two *haploid* daughter cells. The purpose of synapsis should now become clear. The function of the first division is to produce two daughter cells containing one homolog from each pair of homologous chromosomes. Synapsis results in a pairing of the homologs to insure that during anaphase I no two chromosomes from the same pair of homologs will end up in the same daughter cell. Ultimately each gamete should contain only one chromosome from each pair of homologs. The key to that outcome lies here in the first division of meiosis. Because the chromosome number is reduced to haploid in the first division of meiosis, this division is often referred to as the reduction division of meiosis. At this point in spermatogenesis the two products of the first division are called secondary spermatocytes. In ovigenesis, since the division of the cytoplasm is unequal, the first division produces a secondary ovicyte and the first polar body.

2. The second meiotic division is now ready to proceed. The function of the second division is basically to produce two single-stranded chromosomes from each of the doubled chromosomes. In this sense, the second division of meiosis is similar to the single mitotic division. However, it differs from mitosis in several respects, one of which is that the products of this division are not genetically identical to each other or to their parent cell.

As in the first division, the second meiotic division is divided into phases. There is a prophase II and metaphase II that include activities that prepare the cell for the actual division. In anaphase II the actual division of the paired chromosomes takes place, resulting in the formation of two new sets of chromosomes that migrate to opposite poles of the cell in very much the same way as they do in the anaphase of mitosis. Telophase II completes the division process. In spermatogenesis each haploid secondary spermatocyte produces two haploid spermatids. Thus one diploid primary spermatocyte produces four haploid spermatids, none of which are identical to each other nor to any of their ancestral cells. There are two basic reasons why the four spermatids are not identical. First, they contain one of two homologous chromosomes and, generally, homologous chromosomes are not identical in their genetic makeup. Second, the crossing-over phenomenon that occurs in the first division of meiosis results in a further mixing of the genes between homologs. Spermatogenesis and ovigenesis will be explained in more detail later. Meanwhile it should be noted that meiosis is only one step in these two processes.

THE DISTRIBUTION OF GENES FROM GENERATION TO GENERATION

Segregation and Recombination of Genes

A natural law exists that governs the distribution of genes from one generation to the next. This law employs the rules of probability. It may be referred to by anyone of several names. It may be called the law of segregation, or the law of segregation and recombination, or the first law of genetics. Some refer to it as Mendel's First Law in honor of the Austrian monk who discovered it and published his findings in 1865. The whole subject of gene segregation and recombination and their expression in the phenotype is sometimes referred to as Mendelian genetics. Regardless of the terminology used, the law governs two basic events: segregation of genes in the gametes and recombination of genes in the zygote.

Genes segregate in the gametes during the process of meiosis in spermatogenesis and ovigenesis. The two members of each pair of homologous chromosomes are distributed to a separate gamete. Except in rare cases of abnormal meiosis, two homologous chromosomes are never found in the same gamete. Therefore, the two genes found at any two loci on a pair of homologous chromosomes are never found in the same gamete. The genes will segregate in the gametes in all possible combinations due to chance. Individuals of the genotype AA will produce only one kind of gamete, the kind containing one A-gene. In similar fashion the aa genotype will produce only *a*-bearing gametes. From this bit of information, the conclusion can be made that homozygous individuals (homozygotes) produce only one kind of gamete. On the other hand, heterozygotes can produce two kinds of gametes for every pair of heterozygous genes they possess. If we are considering the genes at only one locus, the individual of the genotype *Aa* will produce only two kinds of gametes, those containing an *A*-allele and those containing an *a*-allele.

During the mating process, when the gametes from the male and female unite, the diploid number of chromosomes is once again restored in the zygote. Genes are again recombined into pairs. This recombination process results in one or more possible kinds of genotypes. The genotype that results from the process of recombination depends upon which genes were being carried in the gametes, which in turn was dependent upon the genotypes of the parents.

The Six Basic Crosses

When dealing with any trait in a population that is controlled by two alleles, *A* and *a*, three different genotypes are possible: *AA*, *Aa*, and *aa*. It is possible to combine these three genotypes into the six crosses shown in Table 5-1.

The six crosses can be categorized into three groups based upon the numbers of kinds of progeny produced. The first three crosses are similar in at least two ways: All individuals being mated are

TABLE 5-1.   THE SIX BASIC CROSSES

Genotype of Parents	Genotypes of Progeny
1. $AA \times AA$	All AA
2. $AA \times aa$	All Aa
3. $aa \times aa$	All aa
4. $Aa \times AA$	$\frac{1}{2}Aa : \frac{1}{2}AA$
5. $Aa \times aa$	$\frac{1}{2}Aa : \frac{1}{2}aa$
6. $Aa \times Aa$	$\frac{1}{4}AA : \frac{1}{2}Aa : \frac{1}{4}aa$

homozygous, and only one kind of progeny is produced from each cross. Since homozygotes can produce only one kind of gamete, those gametes can unite as fertilization in only one way. This is shown in the following three illustrations:

Cross number 1

Genotypes of the parents AA × AA

Gametes of the parents   Ⓐ   Ⓐ

Genotype of the progeny AA

Cross number 2

Genotypes of the parents AA × aa

Gametes of the parents   Ⓐ   ⓐ

Genotype of the progeny Aa

Cross number 3

Genotypes of the parents aa × aa

Gametes of the parents   ⓐ   ⓐ

Genotype of the progeny aa

Based upon the results of these three crosses it can be concluded that matings between any two homozygous individuals will produce only one kind of genotype among their progeny.

Two of the crosses, numbers 4 and 5 in Table 5-1, are similar in that one individual is heterozygous and one is homozygous, and that two kinds of progeny are produced from each cross. In addition, the genotypes of the two kinds of progeny are identical to those of the parents as illustrated in cross number 4.

Genotypes of the parents AA × Aa

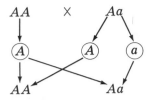

Gametes of the parents

Genotypes of the progeny AA Aa

Based upon the results of crosses 4 and 5 it can be stated that the mating of a heterozygote to a homozygote will produce progeny one-half of which will be heterozygous and one-half of which will be homozygous. The genotypes and genotypic proportions among the progeny will be the same as those of the parents.

Cross number 6 is in a category all its own. It involves a mating of two heterozygous individuals. As has been the case with each of the other five crosses, this cross also illustrates how the genes segregate in the gametes and recombine in the progeny in all possible combinations due to chance.

Genotypes of the parents Aa × Aa

Gametes of the parents A a A a

Genotypes of the progeny AA Aa aA aa

The progeny of this cross are distributed in the ratio of $\frac{1}{4} AA$: $\frac{1}{2} Aa$: $\frac{1}{4} aa$. Stating it in another way, the genotypic ratio of a cross between two heterozygotes is 1 : 2 : 1. This cross is special in that all possible kinds of genotypes that can exist in a system with only two alleles are present among the progeny. For this reason, when studying the ways in which genes express themselves in phenotypes, an attempt should be made to obtain this particular mating.

An individual heterozygous for one pair of genes (Aa) is called a monohybrid. The prefix mono- refers to one pair of genes; the suffix -hybrid refers to the fact that those genes are heterozygous. In genetics the word hybrid means heterozygous or heterozygote. A cross of two individuals that are heterozygous for one pair of genes is called a *monohybrid cross*. Essentially all of the crosses dealing with one pair of alleles will involve one or more of these six crosses. The principles dealing with these crosses will be used later in the solution of crosses involving two or more pairs of genes. Because of the importance attached to the six basic crosses, especially the monohybrid cross, it is recommended that the crosses and their genotypic results be committed to memory.

INTERACTION BETWEEN ALLELES

In dealing with phenotypes and the genotypes that control them we must keep in mind that most of the time the effects of two genes must be considered. If the two genes in any genotype are the same, as in *AA* and *aa*, their effects will be the same upon the phenotype. However, when the heterozygote is involved, we must consider the effects of both of these alleles. Studying the effect of the heterozygous genotype on the phenotype lets us decide how the two alleles interact. If the phenotype of the heterozygote (or hybrid) is identical to one of the homozygotes we say that the gene that is present in that particular homozygote is completely dominant over its allele. The allele in the heterozygote that is not expressed is called recessive. In most situations when the term dominant or dominance is used, it will mean complete dominance unless otherwise specified. Various degrees of allelic interaction other than complete dominance can exist and will be discussed later.

Crossing the Purebreds

Poultry of the White Leghorn breed have what is called a single comb. This comb type is contrasted to the rose comb present in the White Wyandotte breed. The rose comb is short and has a folded appearance that in some way resembles a rose. The rose-comb trait is the result of the presence of a dominant gene which we will symbolize with the letter *R*. Since it is dominant, the rose-comb phenotype can result from either the homozygous (*RR*) or heterozygous (*Rr*) genotypes. The single-comb phenotype will result only from the homozygous state of the recessive allele (*rr*). Purebred White Wyandotte birds are more often homozygous for the rose-comb trait than they are heterozygous. The problem of the heterozygote will be dealt with later. Assuming homozygosity for the dominant phenotype, crosses between the two breeds of poultry

will produce rose-combed progeny that are heterozygous (RR ×
rr = all Rr). This cross is the same as cross number 2 of the six basic
crosses described earlier. The cross between these two purebreds is
called a parental or P_1 cross. The heterozygous progeny are referred
to as the first filial or F_1 generation.

The Monohybrid Cross

A cross between two heterozygous rose-combed birds constitutes the
monohybrid cross. To be sure that the rose-combed birds are indeed
heterozygous, birds should be chosen that have one homozygous
recessive parent as demonstrated in the previous section. The genotypic ratio of the monohybrid cross for these alleles is 1 RR : 2 Rr :
1 rr. The two genotypes having the dominant gene R will express the
rose-comb phenotype. Therefore the theoretical phenotypic ratio of
the monohybrid cross when complete dominance is involved is three
dominant phenotypes to one recessive. In this case, as shown in
Figure 5-4, three rose-combed birds are produced for every single-
combed bird. Slight deviations from this ratio can be expected in
actual crosses. Assume that from several monohybrid matings 100
chicks were produced. According to the theoretical ratio, 75 rose-
combed chicks and 25 single-combed ones would be expected.
However, deviations of up to four or five in each group would not be
unusual in a population of this size.

The Testcross

As we have already seen, when complete dominance is involved
between alleles, a problem exists as to our ability to distinguish between the dominant homozygote (RR) and the heterozygote (Rr).
The heterozygote in this situation is often called a carrier. In situations such as this the genotype R_- can be used to represent both of
the dominant genotypes. The dash indicates that either the dominant
or the recessive allele could be present as the second gene but we do
not know which it is. A breeding test called the testcross can be
performed to help determine which of the two genotypes is present

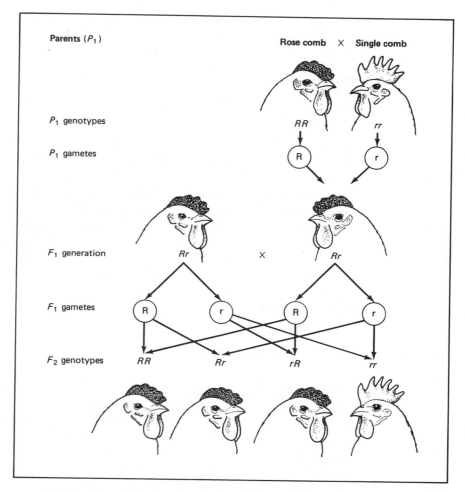

Figure 5-4 Genotypes and phenotypes of F_1 and F_2 generations resulting from crosses between homozygous rose-combed and single-combed chickens.

in the dominant phenotype. Suppose that a Wyandotte breeder is interested in obtaining the services of a new cock (rooster) to breed to his or her hens and does not want to risk the chance of producing some heterozygous chicks. Mating this cock to several single-combed hens would provide the breeder the chance to determine the genotype of the cock before breeding it to the hens. As shown in the following illustration, if the cock were heterozygous, mating it to single-combed hens should produce progeny one-half of which would be single-combed and one-half rose-combed:

Parents: Rose-combed male (Rr) × single-combed female (rr)

Progeny: $\frac{1}{2}$ Rose-combed chicks (Rr) : $\frac{1}{2}$ Single-combed chicks (rr)

The appearance of any single-combed chicks would be genetic proof of heterozygosity in the sire. On the other hand, if the cock were homozygous, no single-combed chicks would be expected. Since the possibility exists that it could be heterozygous and still sire several rose-combed chicks before any single-combed ones appeared, five or ten or more progeny should be observed before deciding that the cock is homozygous. If it is heterozygous, the probability that any one of its progeny would be rose-combed is one-half. The probability that ten progeny would all be rose-combed is $\frac{1}{2}^{10}$ or one chance in 1024. The logical conclusion to make if the cock should sire ten rose-combed progeny and no single-combed progeny is that it is almost surely homozygous.

Intermediate Degrees of Allelic Interaction

Figure 5-5 illustrates graphically several degrees of allelic interaction. It shows that when two alleles are present together in the same geno-type, they do not always interact with one allele exerting complete dominance over the other. The roan color pattern in the Shorthorn breed of cattle illustrates this very well. (Refer to Figure 2-2 in Chapter 2 for a photograph of a roan Shorthorn bull.) Cattle of this breed may express any one of three colors: red, roan or white. Red results from the homozygous state of one allele (RR), white is produced by the opposite allele also in the homozygous state (rr). The roan pattern is the result of the heterozygous condition (Rr). There is no dominance between the two alleles. The roan pattern is made up of both red hairs and white hairs growing amongst each other. The amount of roan varies from a few red hairs on an almost all-white animal to a few white hairs growing on an almost all-red individual. For this reason there is the possibility that some roan cattle may be recorded as white while others may be recorded as red.

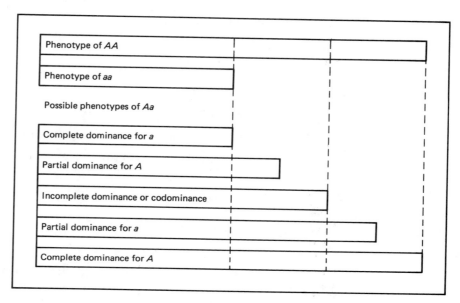

Figure 5-5 Graphic illustration of the various degrees of allelic interaction.

There is no universal agreement as to the terminology that should be used to describe this type of allelic interaction. The terms used most often are incomplete dominance or codominance. Other terms that have been used are lack of dominance, partial dominance, and blending inheritance. Whether the two alleles exert equal dominance upon one another, whether there is a complete absence of dominant effects, or whether there are varying degrees of partial dominance between them, one thing is certain: A separate phenotype exists for each genotype. For the sake of simplicity, we will refer to this type of allelic interaction as blending inheritance. Based upon our knowledge of the six basic crosses presented earlier, the possible matings and their expected results are those indicated in Table 5-2.

Of particular note is the monohybrid cross. The phenotypic ratio of the monohybrid cross with lack of dominance between alleles is 1 : 2 : 1. Note that this phenotypic ratio differs from that obtained from a monohybrid cross involving dominance. Also note that it is identical to the expected genotypic ratio of all monohybrid crosses.

**TABLE 5-2. PHENOTYPIC RESULTS
OF BLENDING INHERITANCE**

Phenotype of Parents	Phenotypes of Progeny
Red (*RR*) × red (*RR*)	All red (*RR*)
White (*rr*) × white (*rr*)	All white (*rr*)
White (*rr*) × red (*RR*)	All roan (*Rr*)
Roan (*Rr*) × red (*RR*)	$\frac{1}{2}$ roan (*Rr*) : $\frac{1}{2}$ red (*RR*)
Roan (*Rr*) × white (*rr*)	$\frac{1}{2}$ roan (*Rr*) : $\frac{1}{2}$ white (*rr*)
Roan (*Rr*) × roan (*Rr*)	$\frac{1}{4}$ red (*RR*) : $\frac{1}{2}$ roan (*Rr*) : $\frac{1}{4}$ white (*rr*)

TWO OR MORE PAIRS OF GENES

In the previous section we dealt with inheritance in which only one allelic system was involved. The principles associated with one pair of genes may be used to determine the results of two-pair crosses. To illustrate how the genes for two sets of independent characters segregate, let us consider a cross between Hereford and Angus breeds of cattle. Angus cattle express the dominant black color (*B_*), while Hereford cattle bear the recessive red (*bb*). The white-faced pattern of the Hereford is controlled by a dominant gene (*H*), while the absence of white on the body is the result of the homozygous state of the recessive allele (*hh*). (Refer again to Figure 2-2 for photographs that illustrate the color patterns in Angus and Hereford cattle.) Crosses between the two breeds produce the familiar black "baldies"—calves that express the dominant black of the Angus and the dominant white-faced pattern of the Hereford. With respect to these two independent sets of characters, the F_1 calves are heterozygous for two sets of alleles and are called *dihybrids*. The assumption is that the purebred parents will in all probability be homozygous. Based upon that assumption the cross can be illustrated as follows:

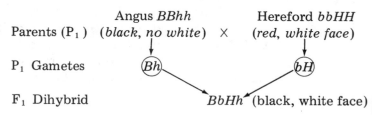

	Angus *BBhh*		Hereford *bbHH*
Parents (P₁)	(*black, no white*)	×	(*red, white face*)
P₁ Gametes	*Bh*		*bH*
F₁ Dihybrid		*BbHh* (black, white face)	

136

The Dihybrid Cross

A cross between two of the heterozygous white-faced, black animals produced from the previous cross constitutes a dihybrid cross. The results of the dihybrid cross can be determined by using the principles developed from the monohybrid cross. The dihybrid cross can be visualized as two simultaneous monohybrid crosses. One of these monohybrid crosses would deal with the genes for black and red: Bb (black) $\times$ Bb (black). As we will recall from the previous section, this cross should produce calves three-fourths of which are black and one-fourth of which are red. The second monohybrid cross would involve the genes for the presence or absence of white on the body: Hh (white face) $\times$ Hh (white face). The product of this cross is also a theoretical 3 : 1 ratio for white face and solid color, respectively. Because the genes for these two sets of traits are inherited independently of each other, the ratio of phenotypes for one trait should exist within each of the phenotypic groups for the other trait. In other words, three-fourths of all black cattle should bear the white face while one-fourth should be solid black; and, three-fourths of all red cattle should bear the white face while one-fourth should be solid red. The results of combining these two sets of traits in a dihybrid cross can be calculated by simply multiplying the two ratios:

(3 Black : 1 Red) $\times$ (3 White face : 1 Solid color) =

   9 White-faced black (the two dominant traits)

   3 White-faced red (one dominant and one recessive trait)

   3 Solid black (one recessive and one dominant trait)

   <u>1</u> Solid red (the two recessive traits)

   16 Total possible combinations

From these results, the following rule can be formulated: The *phenotypic* ratio of a *dihybrid* cross involving two sets of dominant and recessive traits is 9 : 3 : 3 : 1. With respect to the original P_1 mating between black Angus and Hereford cattle, the F_1 is the dihybrid and the F_2 includes the sixteen progeny distributed according to the 9 : 3 : 3 : 1 ratio.

The method of solving the dihybrid cross which has been employed here may be referred to as the algebraic method. Using this method presumes an understanding of monohybrid crosses and ratios. Another method called the square or checkerboard, or Punnitt Square method will be employed in a later problem. A comparison of the two methods will also be made at that time.

The Two-Pair Testcross

As in the case of one-pair crosses involving dominant and recessive genes, the problem exists of not knowing whether the genotypes of individuals bearing a dominant trait are homozygous or heterozygous. Because of this, the genotype of an individual bearing two dominant traits can be expressed using dashes for the unknown genes ($B\_H\_$). A two-factor testcross would involve the mating of this double dominant individual to one or more individuals with the double recessive genotype ($bbhh$). If no red calves were produced from five to ten matings of this type, it could be concluded that the dominant parent is probably homozygous for black. If all the calves had white faces, it could be further concluded that the white-faced parent was probably homozygous for that trait, too. In order for the recessive trait to be expressed, the recessive gene must be present in both of the parents.

Assuming heterozygosity in the dominant parent, let us examine the testcross to determine the nature of the progeny. This cross can also be separated into two one-pair crosses. The cross $Bb \times bb$ will yield a $1 : 1$ ratio of black to red calves regardless of whether they have the white face or not. The cross $Hh \times hh$ will also produce a $1 : 1$ ratio for the traits white face and absence of white regardless of whether the basic color is black or red. The actual results can be calculated as before by multiplying two ratios.

(1 Black : 1 red) $\times$ (1 white face : 1 solid pattern) =

1 White-faced black :

1 White-faced red :

1 Solid black :

1 Solid red

Notice that the same four phenotypes obtained from the dihybrid cross are obtained but in the ratio of 1 : 1 : 1 : 1. The absence of a recessive trait in the testcross progeny would suggest the absence of the recessive gene in the dominant parent.

The Trihybrid Cross

Using the results of the dihybrid cross and considering another monohybrid cross, it is a relatively easy matter to obtain the results of a trihybrid cross. The polled condition (absence of horns) in cattle is caused by a dominant gene P. The presence of horns is controlled by the homozygous state of the recessive allele (pp). A monohybrid cross ($Pp \times Pp$) produces polled and horned progeny in the ratio of 3 : 1, respectively. By multiplying this 3 : 1 ratio times the 9 : 3 : 3 : 1 ratio already obtained among the progeny of the dihybrid cross involving black, white-faced cattle, we obtain the phenoypic ratio among the progeny of a trihybrid cross.

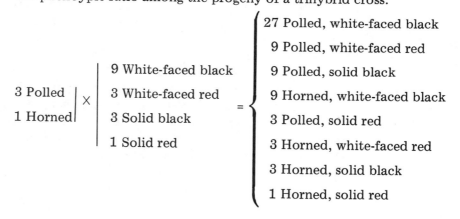

3 Polled
1 Horned
$\times$
9 White-faced black
3 White-faced red
3 Solid black
1 Solid red
=

27 Polled, white-faced black

9 Polled, white-faced red

9 Polled, solid black

9 Horned, white-faced black

3 Polled, solid red

3 Horned, white-faced red

3 Horned, solid black

1 Horned, solid red

Modifications of the 9 : 3 : 3 : 1 Ratio

To illustrate how allelic interaction other than dominance affects the classic 9 : 3 : 3 : 1 ratio, let us consider two traits in horses. The natural trotting gait in horses is apparently controlled by a dominant gene, *T*. The natural pacing gait is the result of the recessive allele when in the homozygous state (*tt*). Because pacing horses are able to run just a bit faster than trotting horses, many natural trotting horses within the Standardbred breed have been trained to pace. For this reason, it is sometimes difficult to determine the natural gait of these horses.

The second set of genes which will be considered determine the palomino color. This popular color in horses is actually a modification of the chestnut color. The genotype *dd* will be used to represent the chestnut color. The allele *D* is called the dilution gene. These alleles do not express any dominance between them so the term blending inheritance or incomplete dominance can be used to describe this relationship. The heterozygous genotype *Dd* results in a partial dilution of the chestnut color resulting in palomino. The homozygous genotype for the dilution gene usually results in an impure white color. The terms pseudoalbino or cremelo are sometimes used to denote this color depending upon the actual appearance of the white. It is not a true albino because the eyes do contain some pigmentation. This type of coloration is somewhat different from the roan trait in cattle in that the palomino color is more of an actual blend of the white and chestnut colors rather than an interspersion of white and chestnut-colored hairs.

Two types of matings can be performed involving homozygotes for both sets of traits. In one case homozygous white pacers (*DDtt*) can be mated to homozygous chestnut trotters (*ddTT*). The product of this cross is the dihybrid *DdTt*. The phenotype will be that of a palomino trotter. The second type of mating would be between a white trotter (*DDTT*) and chestnut pacer (*ddtt*). The results of this cross would be the same as the first.

A mating among horses of the genotype *DdTt* would constitute the dihybrid cross. Many such matings would be necessary to obtain enough foals to illustrate the phenotypic ratio expected from such a cross. In this example, we will use the checkerboard method to solve the dihybrid cross. This method requires more time and allows for a greater possibility for error compared to the algebraic method. However, the checkerboard method illustrates more accurately the process of gene segregation and recombination as it happens naturally.

The first step is to determine the different kinds of gametes that each genotype can produce. Since both members of the cross have the same genotype, they will produce the same kinds of gametes. To make this determination we must recall some of the principles of meiosis. Remember that in any one gamete, only one chromosome from any pair of homologs will be found. This means that for each pair of alleles, only one can appear in a single gamete. Since the two sets of genes are inherited independently of each other, one gene from each set will appear in each gamete. Calculating all the possible combinations due to chance we find that the dihybrid DdTt is expected to produce the four gametes: *DT*, *Dt*, *dT*, and *dt*.

The next step is to construct a chart containing columns and rows designed somewhat like a checkerboard. At the head of each column will be written the four types of gametes from one parent. Down the left side of the chart at the beginning of four rows will be placed the symbols of the gametes from the second parent. Next, in each of the cells or individual boxes in the checkerboard pattern are written genotypes determined by the genes carried in the two gametes listed at the beginning of each intersecting row and column. This process is illustrated in Figure 5-6.

In this example the checkerboard will have 16 individual cells or boxes as will any example involving a dihybrid cross. The number 16 results from the fact that each of the 4 gametes from one parent can combine with any one of the 4 gametes from the second parent. Therefore, all possible combinations due to chance results in a total of 4 × 4 or 16 genotypes. As we examine those 16 genotypes, however, it can be seen that some of them are duplicates of others. In fact, there are only nine different genotypes distributed according to the following ratio:

1 *DDTT*	2 *DdTT*	1 *ddTT*
2 *DDTt*	4 *DdTt*	2 *ddTt*
1 *DDtt*	2 *Ddtt*	1 *ddtt*

$$\frac{1}{4} \quad + \quad \frac{2}{8} \quad + \quad \frac{1}{4} = 16 \quad \text{Total}$$

Notice that the 1 : 2 : 1 genotypic ratios expected from monohybrid crosses are maintained within each set of alleles.

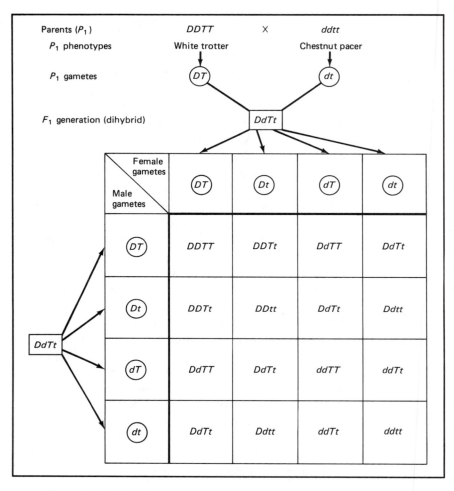

Figure 5-6   The F_1 and F_2 progeny of a mating between homozygous white trotting horses and chestnut pacers.

Finally, the genotypes are translated into phenotypes. The three genotypes at the D-locus produce three separate phenotypes due to the absence of dominance between the alleles: white, palomino, and chestnut in a 1 : 2 : 1 ratio. The genes controlling the type of gait produce trotters and spacers in a 3 : 1 ratio, respectively. The combined results show the following:

3 White trotters

6 Palomino trotters

3 Chestnut trotters

1 White pacer

2 Palomino pacers

1 Chestnut pacer

STUDY QUESTIONS AND EXERCISES

1. Define each of the following terms:

allele

anaphase

carrier

centriole

centromere

character

chromatid

chromosome

codominance

crossing-over

dihybrid

dihybrid cross

diploid

dominant

DNA

F_1 generation

F_2 generation

felial

first meiotic division

gamete

gene

genetics

genotype

haploid

heredity

heterozygous

homolog

homologous

homozygous

hybrid

incomplete dominance

interphase

locus

meiosis

metaphase

mitosis

monohybrid

monohybrid cross

nucleus

P_1 generation

partial dominance

phenotype

polled

progeny

purebred

recessive

recombination

second meiotic division

segregation

sire

somatic cell

spindle

synapsis

telophase

testcross

tetrad

trait

trihybrid cross

2. Why is the study of genetics an important part of the study of domestic animals?

3. Name several domestic animals and tell how many chromosomes are normally present in a single body cell for each.

4. How do genes carry out their function at the cell level?

5. How many different genotypes can exist in a population for a trait controlled by two alleles? Using appropriate symbols, show the genotypes.

6. Explain what happens in a cell as it prepares to undergo mitosis.

7. Compare the daughter cells to the parent cell that has undergone mitosis.

8. Explain the function and significance of mitosis.

9. Why is chromosome duplication in cell division important?

10. Why does the nuclear membrane disappear during mitosis?

11. Where does meiosis take place?

12. Explain the function and significance of the first division of meiosis.

13. Compare the first meiotic division with mitosis.

14. Why is synapsis important?

15. Contrast anaphase I of meiosis to the anaphase of mitosis.

16. Compare the second meiotic division with mitosis.

17. Compare the first and second divisions of meiosis.

18. Explain the significance and function of the second division of meiosis.

19. Why are the daughter cells produced from the process of meiosis not identical to one another?

20. How many different kinds of gametes can be produced by a homozygote?

21. How many different kinds of gametes can be produced by an individual that is heterozygous for one pair of genes?

22. What is the genotypic ratio of a monohybrid cross?

23. Compared to the other five basic crosses, what is significant about the monohybrid cross?

24. What is meant by the statement, "heterozygotes do not breed true"?

25. What is the phenotypic ratio of a monohybrid cross when complete dominance and recessiveness is involved?

26. What is the phenotypic ratio of a monohybrid cross when co-dominance or blending inheritance is involved?

27. What is the significance of the testcross?

28. How can it be determined whether a particular trait is affected by genes which express complete dominance or codominance?

29. Determine the numbers of different kinds of gametes that can be produced from each of these genotypes: *AABB*, *AABb*, *AaBB*, *AaBb*, *aaBB*, *aaBb*, *aabb*.

30. What is the genotypic ratio of all dihybrid crosses?

31. What is the phenotypic ratio of a dihybrid cross involving two traits both controlled by dominant and recessive alleles?

32. What is the theoretical phenotypic ratio of a dihybrid cross involving one trait controlled by complete dominance and a second controlled by blending inheritance?

33. How many different genotypes can be produced from a dihybrid cross?

34. Explain the circumstances that might call for a testcross to be performed?

35. Describe the difference between the palomino color in horses and the roan color in Shorthorn cattle.

36. Compare the checkerboard method of solving two-pair crosses to the algebraic method.

PROBLEMS

1. Most breeds of cats have short hair which is caused by a dominant gene. Breeds of cats with long hair are homozygous for the recessive allele. Assume that a tom with long hair is mated to several females that are assumed to be homozygous for short hair, what genotypes and phenotypes would be presented in the F_1 kittens?

2. Suppose that several matings eventually occurred among the F_1 generation in the above problem and produced a total of 32 F_2 kittens. How many of these kittens should have long hair?

3. How many of the F_2 kittens in problem two should be homozygous for short hair?

4. What procedure should be followed to determine which of the short-haired cats are heterozygous? What is this procedure called?

5. Is it ever possible for long-haired cats mated among themselves to produce any short-haired cats? Explain.

6. The MN blood group system in humans is controlled by two alleles, *M* and *N*, exhibiting codominance. The genotype *MM* produces the blood group M, *NN* produces the blood group N, and the heterozygote produces the group MN. What blood groups would you expect to find among the children from a family where the mother was M and the father was N?

7. Is it possible for a man with M blood to be the father of a child with group N blood? Explain.

8. White guinea pigs mated to yellow guinea pigs produce all cream-colored offspring. Cream guinea pigs mated together usually produce offspring in the ratio of 1 yellow : 2 cream : 1 white. How are these colors inherited?

9. What would you expect from a mating between white and cream-colored guinea pigs?

10. The black color in Hampshire swine is dominant to the red as found in Durocs. The white belt in Hampshires is also a dominant character. What phenotype should be expressed by the pigs from a cross between Hampshires and Durocs assuming homozygosity in both cases?

11. If the F_1 pigs from problem 10 were mated among themselves and several litters were produced, what phenotypes would you expect and in what proportions?

12. Suppose that you decide to develop a breed of red Hampshires. Explain the procedures you would use to carry out the project.

13. Which of the four sets of phenotypes produced in problem 11 would be the easiest to establish as true-breeding traits?

14. Explain why Hampshire hogs will sometimes produce pigs that are not belted.

15. The red, white, and roan colors in Shorthorn cattle are controlled by a pair of codominant alleles. Red and white are caused by the homozygous state of their respective alleles, while roan is due to heterozygosity. Some Shorthorn cattle express the

dominant hornless (polled) condition; others have horns. What would you expect from the cross of a white polled Shorthorn and a red horned Shorthorn?

16. Suppose many matings were made among the dihybrid cattle produced in problem 15. What would you expect in their progeny?

SOLUTIONS TO PROBLEMS

1. All kittens should be heterozygous and possess short hair.

2. Approximately one-fourth, or eight.

3. Approximately one-fourth, or eight.

4. Mate them to long-haired cats (homozygous recessive); any short-haired cat that produces a long haired kitten must be heterozygous. This is a testcross.

5. No. Long-haired cats are homozygous recessive, so no gene for short hair can be transmitted.

6. All children should be MN.

7. No. An *N*-group child would have to obtain an *N*-gene from his father.

8. Two alleles expressing codominance or blending inheritance.

9. One-half white and one-half cream-colored offspring.

10. Black with the white belt.

11. Nine belted blacks : 3 belted reds : 3 solid blacks : 1 solid red.

12. Select belted red hogs for mating; red is recessive, so all offspring should be red. Cull any belted hogs that produce nonbelted pigs. As a further step, belted hogs could be mated to solid red hogs to test the belted ones for the presence of a recessive gene.

13. Red, nonbelted; both are recessive phenotypes, so no problems would exist with hidden genes.

14. Since the belted phenotype is dominant, it is possible for some to carry the recessive nonbelted allele. If two heterozygous belted hogs are mated together they could produce some non-belted pigs.

15. If the polled animal were homozygous, all calves would be polled (carriers) and roan. If the polled animal were heterozygous, half of the calves should be polled roans, and the other half should be horned roans.

16. Three polled red : 6 polled roans : 3 polled white : 1 horned red : 2 horned roan : 1 horned white.

6

Inheritance in Domestic Animals

In the previous chapter we learned that genes are studied in pairs, and that there are at least two alleles affecting each trait. We learned further than an interaction exists between the two alleles such that the phenotype of the heterozygote may or may not be the same as the phenotype of one of the homozygotes. That principle of allelic interaction will also be carried forth in this chapter. However, we will introduce some additional factors that will affect the expression of the genes regardless of the kind of allelic interaction. In addition to being dominant or recessive or partially dominant or codominant, a gene may also have a lethal or detrimental effect. Genes can be expressed differently depending upon the sex of the individual. Genes at one locus may modify the expression of genes at a second locus. Instead of there being just two alleles that affect one trait, there may be three or four or even hundreds. These are the kinds of inheritance that will be described in this chapter.

Following a description of the sex hormones is a discussion of the transmission and effects of the genes carried on these chromosomes.

MULTIPLE ALLELES AND LETHAL GENES

Some traits that are exhibited in animals are the result of interactions between more than two alleles. Occasionally, one of the genes involved may have a lethal effect. In this section we will discuss crosses that involve each of these effects as well as those that combine the two effects with one another and with other traits.

Multiple Alleles

The number of alleles that can exist at any two homologous loci in a single individual is limited to just two since only two loci for those genes are present on two homologous chromosomes. However, in a population there could be many alleles. Remember that a system of alleles includes all the different genes that can exist at the same loci on homologous chomosomes. Any example of an allelic system involving more than two alleles is described as a system of multiple alleles. One such example is the ABO blood group system in humans. Three alleles (A, B, and a) express themselves in a population in the form of four blood groups: A, B, AB, and O. Two types of allelic interaction exist. The alleles A and B are completely dominant to a. The relationship between A and B is one of codominance. The gene A controls the production of antigen A in the blood. The allele B controls the production of antigen B. The recessive allele produces no antigen. In a system of three alleles, six genotypes are possible. Two genotypes result in the production of only antigen A, they are AA and Aa. People who have antigen A in their blood are said to have group A blood. Another two genotypes (BB and Ba) produce group B blood containing only antigen B. People who are heterozygous for both antigen-producing genes carry both antigens in their blood and belong to group AB. Group O blood contains no antigens and is the result of the homozygous recessive genotype aa.

Many traits in domestic animals and humans are known to be controlled by multiple alleles; only a few examples of which are presented here. The number of alleles involved varies from three to several hundred.

Effect of a Lethal Gene

In mice, a pair of alleles determines whether the color will be yellow or some other color (nonyellow). When yellow mice are mated to nonyellow mice, the results are consistent in producing yellow and nonyellow offspring in the ratio of 1 : 1. When yellow mice are mated among themselves they never breed true, but produce yellow and nonyellow progeny in the ratio of 2 : 1 respectively. In addition, litter size from the yellow × yellow crosses is smaller than normal. The yellow color is the result of the heterozygous condition of two alleles (Yy). One of the genes in the homozygous state (yy) results in any color other than yellow. The actual color will be determined by other genes at different loci. The genotype YY results in the death of the embryos at a very early stage of pregnancy. The type of allelic interaction may be described as incomplete dominance or blending inheritance. The genotypic ratio of the monohybrid cross is 1 YY : 2 Yy : 1 yy. Because of the death of one of the homozygotes, the phenotypic ratio comes forth as 2 : 1. Lethal genes may also be completely recessive to their normal alleles. Many examples have been described in various research journals. A large majority of those examples in mammals involve some anatomical defect so severe that fetuses are born dead or die shortly after parturition. Similar examples are cited in poultry which usually result in death of the embryo prior to hatching.

Multiple Alleles with Lethal Effects

An interesting set of traits in domestic foxes involves both multiple alleles and lethal genes. The alleles W, w, and p are involved, none of which are dominant to any of the others. The w- and p-alleles are lethal when in the homozygous state (ww or pp), or in the heterozygous state with one another (wp). The homozygote WW produces silver fur color, the heterozygote Ww produces white-face silver, and the heterozygous Wp produces the popular platinum color. The following six crosses with the phenotypes of their progeny are possible:

154

$WW \times WW$ = all WW (silver)

$WW \times Ww$ = $\frac{1}{2} WW$ (silver) : $\frac{1}{2} Ww$ (white-face silver)

$WW \times Wp$ = $\frac{1}{2} WW$ (silver) : $\frac{1}{2} Wp$ (platinum)

$Ww \times Ww$ = $\frac{1}{3} WW$ (silver) : $\frac{2}{3} Ww$ (white-face silver)

$Ww \times Wp$ = $\frac{1}{3} WW$ (silver) : $\frac{1}{3} Ww$ (white-face silver) : $\frac{1}{3} Wp$
(platinum)

$Wp \times Wp$ = $\frac{1}{3} WW$ (silver) : $\frac{2}{3} Wp$ (platinum)

In each of the last three crosses, one of the lethal genotypes is produced which results in the death of the fetus very early in embryonic development. However, when platinum foxes are mated together, occasionally a pure white pup is produced that dies very shortly after birth. The probable explanation is that the white pup carries the lethal pp genotype, but that it failed to kill the embryo as early as expected.

Dominant Genes and Multiple Alleles

A series of three alleles controls three color patterns in cattle. Two of these patterns were described in the previous chapter. The white face or Hereford pattern is controlled by a dominant gene H. The solid pattern is the result of the recessive allele h. A third allele, s, is recessive to both of the other alleles and produces the spotted pattern common in Holsteins when in the homozygous state (ss). The two genes for black (B) and red (b) are alleles with black being dominant to red. This set of traits was also described earlier. Table 6-1 shows the possible combinations of colors and patterns that can result from these alleles.

The results of one dihybrid cross were presented in the previous chapter ($BbHh \times BbHh$); two others are possible. One, $BbHs \times BbHs$ will produce a theoretical ratio of $\frac{9}{16}$ white face blacks, $\frac{3}{16}$ white face reds, $\frac{3}{16}$ spotted blacks, and $\frac{1}{16}$ spotted reds. A second dihybrid cross, $Bbhs \times Bbhs$, will produce an F_2 ratio of $\frac{9}{16}$ solid blacks, $\frac{3}{16}$ solid reds, $\frac{3}{16}$ spotted blacks, and $\frac{1}{16}$ spotted reds. Three other crosses are possible involving parents with two pairs of heterozygous

TABLE 6-1. GENETIC CONTROL
OF SOME COLOR PATTERNS
IN CATTLE

Genotypes	Phenotypes
B_H_	Black with white face
B_hh, B_hs	Black all over
B_ss	Black with white spots
bbH_	Red with white face
bbhh, bbhs	Red all over
bbss	Red with white spots

genes, but they are not the usual kind of dihybrid cross. As an example, an individual of the genotype *BbHs* can be mated to one that is *Bbhs*. Both of these individuals are dihybrids, but not for the same pairs of genes. This cross does not produce the classical 9 : 3 : 3 : 1 ratio. Instead, the results are as follows:

> 3 Solid black
>
> 6 White face black
>
> 3 Spotted black
>
> 1 Solid red
>
> 2 White face red
>
> 1 Spotted red

Two other similar crosses are possible, *BbHh* × *BbHs* which does produce a 9 : 3 : 3 : 1 ratio; and *BbHh* × *Bbhs* which produces a 3 : 3 : 1 : 1 ratio. Both of these crosses should make good practice problems for the reader.

Dominant Genes and Lethal Genes

Horns in cattle are caused by the homozygous condition of a recessive gene (*pp*). The polled trait is due to the dominant allele in either the homozygous or heterozygous condition (*P_*). The monohybrid cross produces the classic 3 : 1 ratio of polled to horned. A trait in cattle known as comprest dwarfism is the result of the heterozygous condition of a pair of alleles that exhibits no dominance between them (*Cc*). One of the alleles in the homozygous state (*cc*) results in cattle of normal size. The second allele in the homozygous state (*CC*), causes death to the embryo in very early stages of pregnancy. This trait is similar to the lethal condition in yellow mice which was described earlier. The monohybrid cross (dwarf × dwarf) results in a ratio of 2 dwarfs : 1 normal phenotype. The dihybrid cross (*PpCc* × *PpCc*) involves a cross between heterozygous, polled dwarfs. A double homozygous P₁ cross is not possible since the homozygous genotype for the lethal gene is not available for crossing. Therefore, the dihybrid F₁ must be produced from a cross between a polled dwarf (*PPCc*) and a horned animal of normal size (*ppcc*), or between a horned dwarf (*ppCc*) and a polled animal of normal size (*PPcc*). In both cases, the progeny will include two types of individuals; polled dwarfs and polled, normal-sized cattle. In both types of progeny the polled phenotype is heterozygous, so choosing the dihybrid presents no problem.

The results of the dihybrid cross can be determined by multiplying the two monohybrid ratios:

(3 polled : 1 horned) × (2 dwarfs : 1 normal) =

6 polled dwarfs :

3 polled normals :

2 horned dwarfs :

1 polled normal

SEX-RELATED INHERITANCE

Several types of inheritance exist that are affected in some way by sex. When the genes are located on the sex chromosomes, the phenotypic ratios are usually somewhat different from the ratios resulting from crosses with the genes located on the autosomes. Furthermore, the expression of some genes can be modified by the presence of the male or female sex hormones in the body.

Sex Chromosomes

Homologous chromosomes are described as being the same size and shape. In male mammals, one of the several pairs of homologous chromosomes present are not completely homologous. The two members of this pair of seemingly unmatched chromosomes are arbitrarily designated as the X- and Y-chromosomes. Observation of the chromosomes of female mammals has shown that they bear two X-chromosomes and no Y-chromosome. During the process of spermatogenesis in mammals two types of gametes are formed, one bearing the X-chromosome, the other type bearing a Y-chromosome. All female gametes will carry an X-chromosome. The sex of the progeny will be determined by which male gamete fertilizes the female gamete.

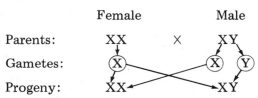

	Female		Male
Parents:	XX	×	XY
Gametes:	Ⓧ		Ⓧ Ⓨ
Progeny:	XX		XY

To avoid some confusion, in birds the letters Z and W may be used to represent the sex chromosomes. The male is ZZ and the female is ZW. Note that in birds, in contrast to mammals, the female has the two sex chromosomes that are different.

Male mammals always obtain a Y-chromosome from their sires, and an X-chromosome from their dams. Female mammals always obtain an X-chromosome from each of their parents. Male mammals always transmit a Y-chromosome to their sons and an X-chromosome to their daughters.

Sex-Linked Genes

Genes that are present on the X- and Y-chromosomes are referred to as sex-linked genes. Those on the X-chromosome are X-linked, those on the Y-chromosome are Y-linked. In mammals, Y-linked genes are present only in males and are passed on only from sire to son. X-linked genes are present in both sexes. In the female mammal, since there are two X-chromosomes, X-linked genes will segregate as do genes on any other pair of chromosomes. Very few traits in domestic animals or humans are known to be controlled by Y-linked genes.

A blood disease in dogs known as hemophilia is controlled by an X-linked recessive gene, h. The dominant allele, H, results in normal clotting of the blood. Hemophilia is a disease that prevents the normal clotting of the blood. In males, only one gene is needed to express either trait because only one X-chromosome is present. Normal males are HY, while males with hemophilia are hY. Normal females can be homozygous (HH) or heterozygous (Hh). Recessive genes do not express themselves in heterozygotes, so these heterozygotes are often called "carriers." Females with hemophilia must have two recessive genes (hh). Hemophilic pups are very fragile and usually bleed to death before reaching maturity. Assuming that only normal dogs are involved in mating, it is possible to produce hemophilic males but no hemophilic females. To produce hemophilic females would require the use of a hemophilic male and carrier female. (See Table 6-2.)

TABLE 6-2. INHERITANCE OF HEMOPHILIC TRAIT IN DOGS

Parents		Progeny	
Sire	Dam	Males	Females
HY (normal)	$\times\ Hh$ (carrier)	$\frac{1}{4}\,HY$ (normal)	$\frac{1}{4}\,HH$ (normal)
		$\frac{1}{4}\,hY$ (hemophilic)	$\frac{1}{4}\,Hh$ (carrier)
hY (hemophilic)	$\times\ Hh$ (carrier	$\frac{1}{4}\,HY$ (normal)	$\frac{1}{4}\,Hh$ (carrier)
		$\frac{1}{4}\,hY$ (hemophilic)	$\frac{1}{4}\,hh$ (hemophilic)

Sex-Influenced Inheritance

A few examples of traits in animals are known where the genes are carried on the autosomes but the phenotypes do not appear in the same proportions in the two sexes. One such example is that of horns in sheep. The gene for horns (H) is dominant in male sheep. Horned rams may be either homozygous (HH) or they may be carriers of the allele for polledness (Hh), which is recessive in the male. For males to be polled, they must be homozygous for the polled gene (*hh*). The same genotypes can occur in females; however, the genes express themselves differently than in males. The polled gene in females is dominant to the allele for horns. This situation does not affect the expression of the two homozygotes; *HH* females grow horns while *hh* females do not, just as in the males. But because of the difference in the allelic interaction in the heterozygote, *Hh* females do not show the horned trait. The difference in the pheno-types of male and female sheep can be illustrated in the progeny from a cross between a horned male (*Hh*) and a polled female (*Hh*), a monohybrid cross (see Table 6-3).

TABLE 6-3.   SEX-INFLUENCED INHERITANCE
OF HORNED TRAIT IN SHEEP

Phenotypes if Male	Genotypes of Progeny	Phenotypes if Female
3 Horned	1 *HH* ⟶	1 Horned
	2 *Hh*	3 Polled
1 Polled ⟵	1 *hh*	

Sex-Limited Inheritance

Sex-limited inheritance refers to any trait that can only be expressed in one of the sexes. There are a number of examples such as milk production in cattle, egg production in chickens, and semen produc-tion by males. These traits, however, are controlled by many pairs of genes. One of the few examples of a sex-limited trait in domestic animals controlled by just one pair of genes is the cock-feathering trait in birds. In most species of birds, including chickens, the male usually develops more colorful and heavier plumage than the female. The cock-feathered trait is expressed only by the male; thus, the designation sex-limited. The hen-feathered trait can be expressed by either sex. The hen-feathered trait is controlled by a dominant gene *H*; the cock-feathered trait is produced by the homozygous state of the recessive allele (*hh*), but only by the male. Table 6-4 shows the genotypes possible and the phenotypes for both sexes.

**TABLE 6-4.   INHERITANCE OF FEATHERING
IN BIRDS**

Genotypes of Either Sex	Phenotypes of Cocks	Phenotypes of Hens
HH	Hen-feathered	Hen-feathered
Hh	Hen-feathered	Hen-feathered
hh	Cock-feathered	Hen-feathered

TWO-PAIR INTERACTIONS

Quite often, two pairs of genes at different loci will work together to produce a certain trait that could not be expressed if the genes were not working together. Two such examples are epistasis and multiple genes. Do not confuse the term multiple genes with multiple alleles.

Epistatic Interaction

Although genes segregate in the gametes and recombine in all possible combinations due to chance in the zygotes at fertilization, sometimes the phenotypes give the appearance that this may not be occurring. One of the causes for this apparent discrepency is a kind of gene action called epistasis. Epistasis involves the interaction between genes that are not alleles—genes at one pair of loci influencing or modifying the expression of genes at a different pair of loci. The genes actually segregate as expected, but epistatic interaction results in modification of the classic 9 : 3 : 3 : 1 phenotype ratio.

A good example of this kind of gene action can be found in cats. Black color in cats is controlled by a dominant gene, B, while brown color is controlled by the homozygous state of the recessive allele (bb). The albino condition is the result of a homozygous pair of recessive genes, aa. The dominant allele in either the homozygous (AA) or heterozygous (Aa) state results in the expression of the black or brown colors. The dihybrid cross ($AabB \times AaBb$) involving two sets of dominant and recessive genes would normally be expected to produce the 9 : 3 : 3 : 1 phenotype ratio. The monohybrid cross involving A-locus genes produces the 3 : 1 ratio of genotypic groups A_- and aa, respectively. The monohybrid cross involving the B-locus genes also produces the 3 : 1 ratio of the genotypic groups B_- and bb, respectively. Multiplying these two monohybrid ratios produces the following results of the dihybrid cross:

9 A_-B_- ———— 9 Black kittens

3 A_-bb ———— 3 Brown kittens

3 aaB_- ⟩ 4 Albino kittens

1 $aabb$

Because the recessive albino genes are epistatic to all color, the two groups of genotypes possessing those recessive albino genes in the homozygous state will express the albino phenotype even though they also possess other genes for black or brown hair. In this example, the usual hair colors are said to be hypostatic to albinism. The word epistasis means to "stand upon," while the word hypostatic means to "stand beneath."

Many other examples of epistasis can be cited that result in other modifications of the 9 : 3 : 3 : 1 ratio. In addition to the 9 : 3 : 4 ratio, various kinds of epistasis result in ratios of 9 : 6 : 1, 9 : 7, 12 : 3 : 1, and 15 : 1.

Multiple Genes

A number of traits in domestic animals are influenced by many genes with many degrees of phenotypic expression. Growth rate and size in animals are two good examples. Since many genes are involved, each gene has a smaller effect than when only one pair of genes is involved. To illustrate how this kind of inheritance works, a rather simple example involving only two pairs of genes can be used. The amount of white on spotted cattle such as Holsteins varies from very little, so that the cattle are almost all black, to a large amount, resulting in cattle that are almost all white. It is thought by some geneticists that this variation is due to the influence of a series of multiple genes. Let us assume that two pairs of alleles are involved, A and a, and E and e. The genes A and E will be thought of as genes that contribute a certain amount of white to the phenotype. The alleles a and e will be thought of as neutral genes and will contribute nothing. The contributing genes are often thought of as additive genes. Further, we will assume that there is no dominance between the alleles and that each additive gene will contribute the same amount of white. The genotype *aaee* will produce the minimum amount of spotting and the genotype *AAEE* will produce the maximum. The five degrees of spotting shown in Table 6-5 will be considered for purposes of this illustration.

TABLE 6-5.   MULTIPLE-GENE INHERITANCE
OF SPOTTING IN CATTLE (HYPOTHETICAL)

Number of Additive Genes	Possible Genotypes	Degree of Spotting
0	*aaee*	10% White, 90% black
1	*Aaee, aaEe*	30% White, 70% black
2	*AAee, AaEe, aaEE*	50% White, 50% black
3	*AAEe, AaEE*	70% White, 30% black
4	*AAEE*	90% White, 10% black

If we further assume that all genes are present in the population with about equal frequency, there should be more cattle with 50 percent white and 50 percent black than with any other phenotype since there are more genotypes for that degree of spotting. To determine the theoretical number of each phenotype that should be present in a random breeding population, let us look at the results of the dihybrid cross. Dihybrid individuals (*AaEe*) can be produced by mating cattle that are almost all black with those that are almost all white. These cattle would most likely be homozygous for opposite alleles (*AAEE* × *aaee* = *AaEe*). The genotypic results of the dihybrid cross will be the same for all such crosses.

$$\frac{\begin{matrix} 1\ AAEE \\ 2\ AAEe \\ 1\ AAee \end{matrix}}{4} + \frac{\begin{matrix} 2\ AaEE \\ 4\ AaEe \\ 2\ Aaee \end{matrix}}{8} + \frac{\begin{matrix} 1\ aaEE \\ 2\ aaEe \\ 1\ aaee \end{matrix}}{4} = 16\ \text{Total}$$

The number of phenotypes are controlled by how many additive genes are present in each of the above genotypes. There are five possible phenotypes existing in the ratios shown in Table 6-6.

TABLE 6-6.   RATIOS OF MULTIPLE-GENE INHERITANCE OF SPOTTING IN CATTLE (HYPOTHETICAL)

Number of Individuals	Genotypes and Their Ratios	Number of Additive Genes	Phenotypes
1	1 *aaee*	0	10% White
4	2 *Aaee*, 2 *aaEe*	1	30% White
6	1 *AAee*, 1 *aaEE*, 4 *AaEe*	2	50% White
4	2 *AAEe*, 2 *AaEE*	3	70% White
1	1 *AAEE*	4	90% White

This problem represents a simple illustration of the type of inheritance that affects many of the so-called economically important characteristics in domestic animals. Fertility, growth rate, feed efficiency, carcass quality in meat animals, egg production in poultry, milk production in dairy cattle, racing ability in horses, and hunting instinct in dogs are some examples of economically important traits in domestic animals. Most traits of this type are affected by many more than just two pairs of genes, but many of the principles illus-

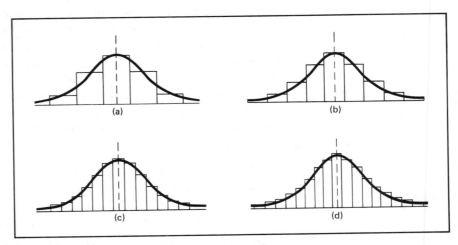

Figure 6-1   Histograms and curves of normal distribution for four populations. (a) Two pairs of genes. (b) Four pairs of genes. (c) Six pairs of genes. (d) Eight pairs of genes.

trated in this two-pair cross apply. There are many, many more phenotypes than this illustration shows, but they tend to fit into a similar type of distribution. The phenotypes of a large part of the population will tend to be near the average, with fewer numbers of animals in the phenotypic groups that are less like the average. In Figure 6-1 the four bar graphs illustrate the distributions of four hypothetical populations controlled by different numbers of additive genes. Graph A illustrates the number of individuals and phenotypic categories for a trait controlled by two pairs of genes similar to the illustration about the degree of spotting in cattle. Graphs B, C, and D illustrate the relative numbers of individuals within phenotypic groups controlled by increasingly greater numbers of additive genes.

STUDY QUESTIONS AND EXERCISES

1. Define each of the following terms:

   autosomes   multiple genes
   epistasis   sex chromosomes
   hemophilia   sex-influenced trait
   hypostatic   sex-limited trait
   lethal   sex linkage
   multiple alleles

2. What is the theoretical maximum number of alleles that can exist for one trait within a population?

3. Explain how sex chromosomes differ in male and female mammals and birds.

4. How is sex determined in domestic animals?

5. What is the minimum number of pairs of genes needed to explain epistasis?

6. Compare the phenotypic ratio of a dihybrid cross involving dominant genes with one involving multiple genes.

PROBLEMS

1. Platinum color in foxes is due to the heterozygous state of two alleles. One allele in the homozygous state produces the silver color, the other allele in the homozygous state is lethal, usually killing the animal during embryonic development. Could a breed of platinum foxes ever be developed?

2. What phenotypic results would you expect from the mating of platinum foxes among themselves?

3. Sometimes the cross in problem 2 produces a pure white pup that usually dies shortly after birth. How can the appearance of the white pup be explained?

4. A third allele also exists at the same locus as the silver and platinum genes in foxes. This third allele when present in the homozygous state or when heterozygous with the platinum lethal gene results in the embryonic death. In the heterozygous state with the silver gene it results in the color called white-face silver. How many different genotypes and phenotypes would you expect from a cross between platinum and white-face silver foxes? Let w and p be the two lethal genes, and W the gene for silver.

5. Could a breed of silver foxes be established?

6. Two alles in cats control the black and yellow colors. These alleles are linked to the X-chromosome and exhibit no dominance between them. In females black is homozygous and yellow in homozygous. The black and yellow genes exert codominant effects in the heterozygote and produce the tortoise-shell color. What colors can exist in male cats?

7. What results would you expect from mating a yellow male to a tortoise-shell female?

8. Comprest dwarfism in cattle is caused by the heterozygous state of two alleles. One allele in the homozygous state produces cattle that develop to a normal adult size. The other allele in the homozygous state is lethal usually causing embyronic death. What would you expect from crosses between red, horned dwarfs and roan, horned cattle of normal size?

SOLUTIONS TO PROBLEMS

1. No. Platinum is due to heterozygosity and heterozygotes do not breed true.

2. One-third of the offspring should be silver, and two-thirds platinum.

3. It is probably a pup that is homozygous for the lethal gene, but was able to survive until birth.

4. The cross would be Wp (platinum) $\times$ Ww (white-face silver). Four genotypes should be produced: pW would die embryonically; the others would be Wp (platinum), Ww (white-face silver), and WW (silver).

5. Yes. Silver (WW) is homozygous and so will breed true.

6. Only black or yellow. Since they have only one X-chromosome both alleles cannot exist and so males normally cannot show the tortoise-shell color.

7. Tortoise-shell and yellow females; black and yellow males.

8. All horned cattle in a ratio of 1 red normal : 1 roan normal : 1 red dwarf : 1 roan dwarf.

7

Genetics
of Animal Breeding

The primary goal of the animal breeder is to develop genetically superior animals. This can be done by identifying those animals possessing superior genotypes for use in breeding programs. In explaining how the breeder may accomplish this, we will discuss the principles of inbreeding, outbreeding, genetic selection, heritability, and hybrid vigor. Several examples and problems will be presented to illustrate how these breeding tools can be used.

GENETIC SELECTION

Genetic selection involves those forces that determine which animals will be allowed to produce the next generation. Animals that are not permitted to reproduce are culled. Culling is a part of the selection process. Most geneticists describe two types of selection: natural and artificial. If the forces that allow certain animals to reproduce are not controlled by humans, natural selection takes place. If the selection forces are controlled by humans it is called artificial selection. These terms seem to imply that controlled selection is not

170

natural or that human beings are not a part of nature. On the contrary, we are as much a part of nature as any other living or nonliving thing.

The type of selection practiced by humans usually involves allowing the better animals, or, hopefully, the best animals, to reproduce and produce the next generation. A very simple definition of selection can be formulated based upon the assumption that the best animals will be chosen: artificial genetic selection can be defined as "mating the best to the best."

Quantitative Versus Qualitative Inheritance

Qualitative traits are those involving phenotypes that are easily distinguished one from another and that are controlled by only few pairs of genes. It is not difficult to distinguish red from black in cattle, or to recognize that a chicken has a rose comb or a single comb. The difference can usually be determined by observation without the aid of some measuring device. Qualitative traits are usually controlled by dominance and recessiveness, blending inheritance, or epistasis. The term *nonadditive* is used to apply to these kinds of gene action. Mating the best to the best for qualitative traits involves mating those that possess the desired trait and culling those that do not. When dominant or epistatic genes are involved, there is always the possibility that recessive or hypostatic genes may be present and not be shown. In this case, breeding tests must be performed in an attempt to uncover these hidden genes. The testcross, which was described earlier, is one such breeding test that can be used for this purpose.

Quantitative traits are those that are controlled by a large number of genes. Each gene usually has only a small effect and there is usually very little, if any, dominance involved within pairs of alleles. The term *additive gene action* is used to describe the kind of genetic control associated with quantitative traits. The discussion in Chapter 6 of degrees of spotting in Holstein cattle provided a simplified example of how quantitative inheritance works. So many genes are usually involved, producing so many phenotypes, that some kind of measuring tool is usually needed to distinguish among the phenotypes. For example, cattle in the feedlot must be weighed periodically to determine their growth rate. The fastest gaining cattle can usually be distinguished from the slowest gaining ones, but even then, the actual amount of gain cannot be determined by casual observation. Most of the economically important traits described earlier are controlled to a large extent by quantitative inheritance.

Mating the best to the best involves mating those animals possessing the desired phenotypes and culling those that do not. The problems associated with knowing which animals are truly best are somewhat different from those associated with qualitative traits. While quantitative traits are controlled to a large extent by additive genes, both environment and nonadditive genes also have some effect. The amount of these other effects varies with the trait.

Hereditary and Environmental Variation

Variation in genetics refers to the differences observed among the phenotypes for any given trait. It is usually measured as deviation from the mean or average for a given population. The best or better animals used in a selection program (mating the best to the best) are those that are superior to the population mean. Differences among animals are caused by two major factors: heredity and environment. Environmental factors include such things as nutrition, weather, disease, and husbandry. Heredity is usually subdivided into additive and nonadditive effects. For the purposes of genetic selection for improvement in quantitative traits, three sources of variation are commonly noted: additive effects, nonadditive effects, and environmental effects.

Heritability and Its Uses

When superior animals are chosen in an attempt to produce improved progeny, it must be taken into consideration that some of their superiority is caused by the environment. That portion of the superiority due to environmental effects cannot be transmitted by way of the gametes. Only that part of the variation due to genetic effects can be expected to be transmitted. Statistical geneticists have devised a method of estimating the proportion of the total variation that can be expected to transfer to the progeny. This ratio is called the heritability estimate. By knowing the heritability estimate for any given trait, breeders can have an idea of how much progress they can make by mating the best to the best. Heritability estimates have been determined for essentially every economically important trait in domestic animals. Table 7-1 lists some examples of heritability estimates for some of the more common economically important traits. Generally speaking, more progress can be made from selection programs for traits with the higher heritabilities than can be made for traits with lower heritabilities.

Suppose that in a large herd of swine, the average litter size at birth is 8 pigs. In an attempt to improve upon this average in the next generation, suppose that the breeder chooses replacement gilts and boars from litters that averaged 12 pigs. The variation or superiority is 4 pigs (12 – 8 = 4). This superiority or variation is also called the selection differential. The question arises as to how much of this 4-pig superiority can be expected to show up in the litters of these selected boars and gilts. The way to find out is to perform the matings and count the pigs. However, before the matings are made we should be able to predict with a reasonable degree of accuracy what the new litter size should be. The heri-

TABLE 7-1. HERITABILITY ESTIMATES
FOR IMPORTANT TRAITS

Trait	Percent Heritability (approximate)
Number born or hatched (fertility)	5–10
Weaning weights in cattle and sheep	30–35
Postweaning growth rate in swine	25–30
Feed efficiency in swine	25–30
Milk production in dairy cattle	20–30
Egg production in chickens	35
Feed efficiency in beef cattle	35–40
Carcass traits in swine and beef cattle	40–60
Postweaning growth rate in cattle and sheep	45–55

tability of fertility is about 10 percent. This means that only about 10 percent of the 4-pig superiority of the selected parents can be expected to be transmitted. Ten percent of four is 0.4; the litters of these selected boars and gilts should only be 4/10 of a pig better than the 8-pig average. That does not appear to be very much progress, and indeed it is not. Traits with low estimates of heritability do not show much progress from selection.

Fortunately, there are at least two other ways that improvement in fertility can be obtained. One of these is to improve the environmental factors such as nutrition and health. Improvement of environmental factors will result in faster progress than attempts to improve genetic factors. Whereas improvement through hereditary means requires at least enough time to turn one generation, improvement through environmental means can often be seen in a matter of days. The other means of improvement in traits with low heritabilities is through the use of outbreeding, which will be explained later in this chapter.

As a second example of how traits having larger estimates of heritability can be improved by genetic selection let us use feedlot rate of gain in beef cattle. The heritability of this trait is approximately 55 percent, which is considered high. Suppose that the adjusted average daily gain of a large population of heifers and bulls in the feedlot is 2.5 lb. The adjustments would be for differences due to sex since bulls tend to gain faster than heifers. Assume that selected bulls gained an average of 3.6 lb per day and heifers an average of 3.0 lb per day. The adjusted gains of heifers are lower than those of the bulls due to the fact that several more females than males are needed in a good selection program; and the greater the number of animals selected, the smaller their average superiority will be. Regardless of the relative numbers of males and females chosen, each group constitutes one-half of the genetic makeup of the progeny; therefore the average gain of the selected cattle is 3.3 lb per day [(3.6 + 3.0) ÷ 2]. Their superiority or selection differential is 0.8 lb (3.3 - 2.5 = 0.8). The predicted amount of this variation that would be expected to be transmitted to the progeny is calculated by multiplying the heritability estimate by the selection differential (0.55 × 0.8 = 0.44). The progeny should gain 0.44 lb per day more than the population average of 2.5 lb per day. Their actual daily gain in the feedlot should be about 2.94 lb (2.50 + 0.44 = 2.94). Most observers would say that this is reasonably good progress, especially when compared to the results of the previous example in swine. It illustrates further that the higher the heritability for a trait, the greater the progress from genetic selection.

MATING SYSTEMS

Once breeders have chosen superior animals to be mated in hopes of producing improved progeny, they then have a choice of using either of two types of mating systems: inbreeding or outbreeding. The difference between these two types of mating systems is basically one of the genetic relationship between the animals being mated. So, before we discuss the mating systems themselves, the subject of relationship should be discussed.

Relationship

Relationship refers to the proportion of genes that two individuals have in common because they are members of the same family. The degree of relationship between two individuals varies depending upon their respective position in the pedigree or on the "family tree." Table 7-2 shows the percentage of genes that any individual has in common with certain other family members over the average of the population. It should be noted that the closer the family relationship, the greater the percentage of genes that any two individuals will have in common.

Two types of family relationships can be described. Collateral relatives are those that are related because they have one or more common ancestors. Full sibs (combined abbreviation of sister and brother) have two common ancestors—they both have the same two parents. The genes that full sibs have in common were obtained from their sire (father) and dam (mother). Incidently, the relationships being described in this section apply to human families as well as to families in domestic animals. An individual's collateral relatives would include sibs, cousins, aunts, uncles, nieces, and nephews. This is true because they have at least one common ancestor.

A second type of relationship is called direct relationship. Direct relatives have a proportion of their genes in common because

TABLE 7-2.   COEFFICIENTS OF RELATIONSHIP BETWEEN AN INDIVIDUAL AND VARIOUS FAMILY MEMBERS

Family Member	Coefficient of Relationship
Identical twin	100
Fraternal twin	50
Full brother or sister (full sib)	50
Father or mother (sire or dam)	50
Child (son or daughter)	50
Half brother or sister (half-sib)	25
Grandparent	25
Great-grandparent	12.5
First cousin	12.5
Great-great-grandparent	6.25
Second cousin	3.125
Third cousin	0.78

one is an ancestor of the other, or one is a descendant of the other. An individual is a direct relative of his or her parents, grandparents, great-grandparents, and so forth. Note that the further back an ancestor appears in an individual's pedigree, the lower the relationship that exists between them and thus the lower the percentage of genes they have in common. See the pedigree and arrow diagram in Figure 7-1. An arrow diagram is a modified pedigree. Only those individuals who are in a direct line from the common ancestor and the first individual in the pedigree are listed. No individual is listed more than one time in an arrow diagram. The arrows indicate to whom genes have been contributed.

Inbreeding

Inbreeding is the mating of animals that are more closely related than the average for the population. As a rule of thumb, inbreeding usually involves individuals that are at least as closely related as

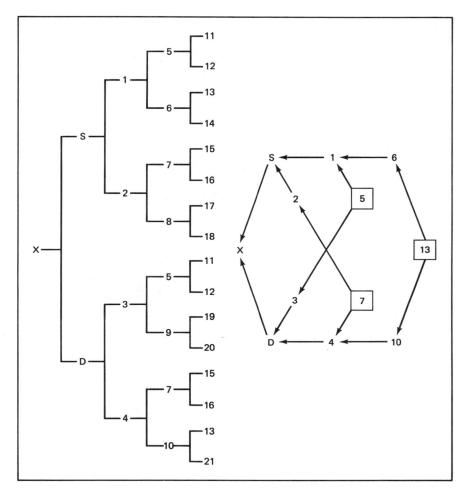

Figure 7-1 A pedigree and arrow diagram illustrating inbreeding.

second cousins, that is, having a minimum coefficient of relationship of 3.125. Figure 7-1 shows the pedigree of an individual whose sire and dam have three common ancestors. Figure 7-2 shows another pedigree with only one common ancestor that appears four times in the sire's pedigree and once in the pedigree of the dam. This second pedigree illustrates a kind of inbreeding called linebreeding. With linebreeding an attempt is made to concentrate the genes of a particular ancestor into the pedigree of an individual.

The genetic effect of inbreeding is to increase the number of

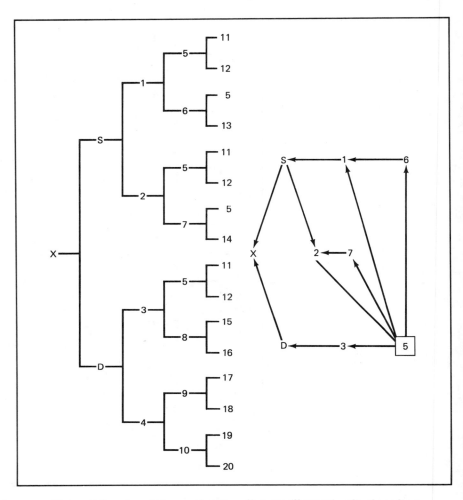

Figure 7-2   A pedigree and arrow diagram illustrating linebreeding.

pairs of genes that exist in the homozygous state. Related animals tend to possess more of the same genes. They are thus more likely to pass the same genes on to their offspring, resulting in more homozygous pairs of genes in those offspring. The increase in homozygosity due to inbreeding occurs with all kinds of genes— additive and nonadditive, and good and bad genes. Inbreeding does not create new genes nor destroy any; it merely rearranges genotypes that were heterozygous and makes more of them homozygous. The degree to which homozygosity is increased due to inbreeding depends primarily upon the closeness of relationship. The closer the relationship, the greater the increase in homozygosity.

Phenotypically, inbreeding usually results in a decline in vigor and performance. When genes become more homozygous, some detrimental recessive genes that had been previously hidden show up in the homozygous state. As we shall explain later, heterozygosity results in what is called hybrid vigor. Inbreeding reduces the number of pairs of heterozygous genes, thus reducing hybrid vigor. Lest it should sound as if inbreeding is all bad, let us look at the bright side of inbreeding. Along with a good selection program, inbreeding can result in the production of high-quality, prepotent breeding animals. Not only are the bad genes rearranged into the homozygous state, the good genes experience the same treatment. Culling the poor animals and allowing the better ones to continue inbreeding can eventually result in a line of animals that are highly homozygous for more of the good genes. These highly homozygous, prepotent animals can then be used in crosses to increase hybrid vigor. Inbreeding is useful in uncovering detrimental recessive genes so that they may be culled. It can also be used to help establish new family lines. For whatever purpose inbreeding may be used, it should always be accompanied by a good selection program.

Traits affected most by inbreeding are those that have the lowest heritability estimates. For example, fertility usually suffers most with an inbreeding program. This is particularly true during the first few generations of inbreeding.

Outbreeding

Outbreeding is the mating of animals less closely related than the average of the population regardless of whether they are members of the same breed or not. As a rule of thumb, if two individuals do not have any ancestors in common for five or six generations back in their respective pedigrees, they are usually thought of as not being any more related than the average of the population. To illustrate, assume that in the two pedigrees in Figures 7-1 and 7-2 individual number 11 at the top of the fifth generation in both pedigrees is the same and is the only ancestor common to both pedigrees. If this were true, the coefficient of relationship between individuals X and Y would be only 0.0039 or approximately 4/10 of 1 percent, very little indeed.

The genetic effect of outbreeding is the opposite of that of inbreeding: It causes more pairs of genes to be heterozygous. Genes are not created or destroyed, only rearranged to reduce the number of pairs in the homozygous state and increase the number of pairs existing in the heterozygous state. Phenotypically, heterozygous animals tend to be more vigorous than homozygous animals. Any inferior or detrimental genes that may have existed in the parents in the homozygous state are more likely to be heterozygous in the outbred individual and thus their effects would not be expressed. Another cause of the vigor observed in outbred animals is apparently due to a rather mysterious interaction between alleles in the heterozygote. From this situation comes the term "hybrid vigor" or "heterosis."

The degree or amount of hybrid vigor observed from outbreeding is affected by two major factors: the degree of "nonrelationship" and the size of the heritability estimate of the particular trait. To some degree, all animals within a given species are related— it is a matter of how far we must go back in their pedigree to find a common ancestor. Fortunately, we can get a general idea of the degree of genetic dissimilarity (nonrelationship) between two individuals without tracing their pedigrees back 30-odd generations or so. It is possible to find relatively unrelated animals within the same breed. Mating animals of this type together would constitute a kind of outbreeding called *outcrossing* or *linecrossing*. Purebred breeders can make use of this kind of outbreeding to obtain a small amount of hybrid vigor to counteract some of the depression in vigor that may have occurred from inbreeding. In some of the larger breeds, it may be possible to obtain individuals for linecrossing that have no common ancestors for at least 10 generations or more.

The genetic dissimilarity is even greater in animals from two different breeds. Outbreeding of this type is the best known and is called *crossbreeding*. Because the genetic makeup is more diverse, the amount of hybrid vigor obtained from crossbreeding is usually greater for a given trait than it is from outcrossing. Genetic diversity in outbreeding can be carried to an even greater extreme with species crosses, of which mules are the classic example. A mule is the product of a cross between a mare (female horse) and a jack (male donkey). The reciprocal cross—between a stallion and a jenny—is possible but the product is called a hinny rather than a mule. Species crosses have limited applications due to the fact that most of them, like the mule, are sterile. Some others are not completely sterile but are highly infertile.

Another major factor affecting the degree of hybrid vigor or heterosis that is obtained from outbreeding is the size of the heritability estimate for the particular trait involved. Traits with lower heritabilities tend to exhibit more heterosis from outbreeding than do traits with higher heritabilities. Fertility tends to reap the greatest benefits from outbreeding compared to the other traits of economic importance because it generally has the lowest estimates of heritability in all species. On the other hand, the quality of the meat from crossbred cattle is only very slightly if any higher than that from purebred cattle. This is due mainly to the fact that the heritabilities of carcass traits is generally high. As explained earlier, traits with the higher heritability estimates are effected to a large extent by additive genes. Interactions among additive genes seem to produce very little hybrid vigor.

Benefits of Crossbreeding

The benefits of crossbreeding generally apply to commercial swine and beef cattle breeders although there are definite benefits in sheep and poultry. Further, the outcrossing sometimes practiced by pure-bred breeders can be thought of as crossbreeding within a breed. Crossbreeding is generally not practiced in dairy cattle for the pur-poses of increasing milk production because the crossbreds usually cannot exceed the production of the best of the Holstein cattle now available. Our discussion here will maily emphasize the benefits of crossbreeding in commercial swine and beef cattle programs, particularly the feeder pig and the cow-calf programs.

No crossbreeding program can be any better than the quality of the animals used to begin the program. Inbreeding has a place in providing high quality, highly homozygous, and thus, prepotent seed stock. Selection is always important in maintaining a high level of performance. The overall management skills of the breeder are of utmost importance, particularly those skills related to nutrition and a good health program.

The first benefits of crossbreeding will be noticed when the first-cross offspring are born. The first-cross offspring are those produced from the mating of purebred parents of different breeds. The benefits will be seen in larger litters, not only in swine but in all litter-bearing animals including dogs, cats, and rabbits. More calves will be born per 100 cows and there will be an increase in the foaling and lambing rates of horses and sheep, as well as a greater percent hatchability in birds. Because of the increased heterozygosity within a crossbred animal, it will be better able to survive during the period between fertilization and parturition or hatching. The net result is more live births per female bred. Additional benefits will be observed as the young animals grow and develop to weaning age. They will grow faster and be less susceptible to stresses of all kinds. The net result is a greater number of heavier animals weaned because they are crossbred.

In addition to the benefits of crossbreeding in the young pig or calf, crossbred dams have even a greater contribution to make. They ovulate more ova, provide a better uterine environment, produce more milk, and exhibit greater "mothering" instinct. The ultimate result of these factors is an even greater number of heavier calves and pigs at weaning. The question arises as to how to maintain a certain level of heterosis in the dams as well as in the offspring. Several crossbreeding programs can be designed to do this, one of which is the three-breed rotational cross.

The program should begin with the best quality animals that can be obtained within the economic limits of the breeder. The breed from which the parents are chosen is not as important as the overall goodness of the individuals chosen. Suppose that purebred Yorkshire gilts are mated to purebred Hampshire boars. The Yorkshire-Hampshire pigs will have the benefits of being crossbred. Replacement gilts will be chosen from these litters and bred to a purebred boar of a third breed, possibly a Duroc. Not only will the pigs be crossbred (50 percent Duroc, 25 percent Hampshire, 25 percent Yorkshire), but so will their dams (50 percent Hampshire, 50 percent Yorkshire). The maximum benefits of heterosis are possible in this cross. Replacement gilts kept from the three-way cross could be bred to a purebred boar of a fourth breed if one were available. Usually this does not result in any more heterosis than choosing a purebred boar from one of the two breeds that are represented in the gilts in the smallest amount. If a Yorkshire boar were chosen, its progeny would be bred to a Hampshire boar and the rotation would continue among those three breeds. After several generations of rotational crossing, the percentage of the three breeds would be about 57 percent of one, 28 percent of the second, and 14 percent of the third. These percentages would vary somewhat from generation to generation. The choice of the breed from which to choose the next sire would be the breed represented in the gilts in the least amount. Figure 7-3 illustrates a three-breed rotational crossbreeding program in swine using Yorkshires, Durocs, and Hampshires as examples.

The benefits of crossbreeding can be measured mathematically. Table 7-3 shows the approximate advantages of the two-breed and three-breed cross over that of their purebred parents for several economically important traits in domestic animals expressed as percent heterosis.

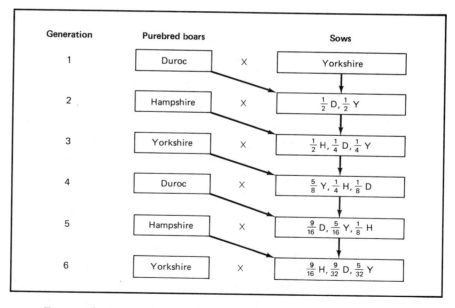

Figure 7-3 Example of a three-breed rotational crossbreeding program in swine.

TABLE 7-3.   ADVANTAGES
OF TWO- AND THREE-BREED CROSSES
OVER
PUREBREDS (PERCENT HETEROSIS)

Trait	Two-Breed Cross	Three-Breed Cross
Percent calving failure	15	25
Litter size at birth	5	17
Litter size at weaning	20	50
Calf weight at weaning	8	20
Litter weight at weaning	35	65

It should be noted that the benefits of crossbreeding cannot be transmitted by way of the gametes. The only way to obtain heterosis is to continue to mate unrelated animals to establish a degree of heterozygosity in the progeny.

STUDY QUESTIONS AND EXERCISES

1. Define each of the following terms:

   ancestor   linecrossing
   breeding value   nonadditive genes
   collateral relative   outbreeding
   crossbreeding   outcrossing
   culling   pedigree
   descendant   progeny testing
   environment   purebreeding
   half-sib   qualitative traits
   heritability   quantitative traits
   heterosis   relationship
   hybrid vigor   selection
   inbreeding   selection differential
   linebreeding   variation

2. Explain how a breeder might design a selection program for a recessive trait? For a dominant trait?

3. What is an economically important trait in domestic animals?

4. List some examples of sources of environmental variation.

5. Name two types of genetic variation.

6. How does the size of the heritability estimate affect progress in a selection program?

7. Cite some examples of traits with low, medium, and high heritability estimates.

8. What kind of a program could you suggest to improve litter size in swine?

9. Why do males generally have a larger selection differential than females?

10. What is the usual genetic effect of inbreeding?

11. What is the usual genetic effect of outbreeding?

12. What is the general phenotypic effect of inbreeding?

13. What is the general phenotypic effect of outbreeding?

14. How does heritability affect the phenotypic effects of crossbreeding? Of inbreeding?

15. How does relationship affect the phenotypic effect of inbreeding? Of outbreeding?

16. How can inbreeding be useful in the improvement of domestic animals?

17. How can outbreeding be useful in the improvement of domestic animals?

PROBLEMS

1. Suppose that in a small herd of beef cattle ten heifers were produced with the following weaning weights measured in pounds: 520, 500, 450, 440, 420, 410, 400, 380, 340, 290. What is the mean weight of this group of heifers?

2. Suppose that the two best heifers were kept for replacements. By how many pounds are they superior to the rest of the group of heifers?

3. If the heritability estimate for this trait were 20 percent, how much of the superiority from the two best heifers would you expect to be transmitted if they were bred to a bull of equal quality?

4. Suppose the three best heifers were kept for replacements and the balance sold. If the three best heifers were bred to a bull that weighed 570 pounds at weaning, what is the selection differential for these selected cattle? Assume the population mean is that calculated in problem 1.

5. If the heritability of weaning weight were 25 percent, what should the calves from the three heifers in problem 4 weigh when they reach weaning age? Include no other factors except those considered in these problems.

SOLUTIONS TO PROBLEMS

1. 415 lb.

2. 95 lb (510 – 415).

3. 19 lb (0.20 × 95).

4. The mean of the selected heifers and bull is [(520 + 500 + 450)/3 + 570] ÷ 2 = 530 lb; the selection differential is 530 – 415 = 115 lb.

5. 444 lb (115 × 0.25 + 415).

8

The Female Reproductive System

The female reproductive system in mammals consists of the two ovaries and the reproductive tract. The reproductive tract consists of the two Fallopian tubes, the uterus, and the vagina. The ovaries and the tract are suspended in the rear part of the abdominal cavity and the pelvic cavity by a hammocklike structure called the broad ligament. Figure 8-1 shows the comparative structure of the reproductive organs of the cow and the sow.

THE FEMALE REPRODUCTIVE TRACT

The Fallopian Tubes

The Fallopian tubes lead from the ovaries to the uterus. They are supported in the part of the broad ligament known as the mesosalpinx. The Fallopian tubes may also be referred to as oviducts, although this term is used more frequently to refer to the reproductive tract of birds. The part of a Fallopian tube nearest the ovary is enlarged into a funnel-shaped structure called the infundibulum. The

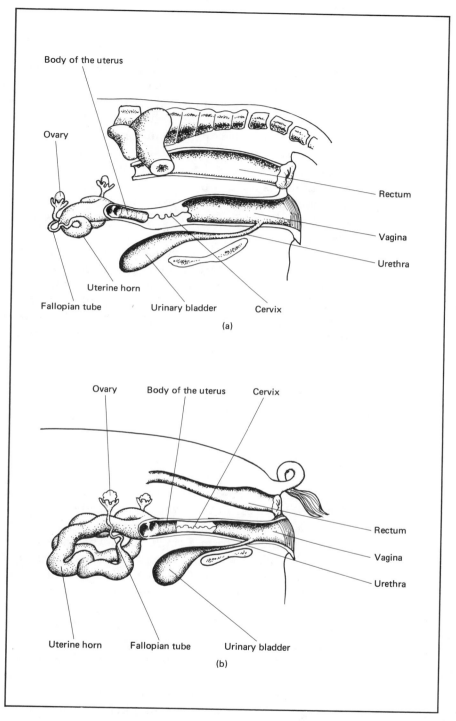

Body of the uterus

Ovary

Rectum

Vagina

Urethra

Uterine horn

Fallopian tube

Urinary bladder

Cervix

(a)

Ovary

Body of the uterus

Cervix

Rectum

Vagina

Urethra

Uterine horn

Fallopian tube

Urinary bladder

(b)

Figure 8-1　Reproductive tracts of (a) a cow and (b) a sow.

infundibulum partially surrounds the ovary, especially during the heat, or estrus, period. Its function is to catch the ovum when it is released from the ovarian follicle at the time of ovulation.

The remainder of the Fallopian tube is divided about equally into two parts called the ampulla and the isthmus. The isthmus connects to the uterus. The diameter of the ampulla will vary some among species but will average about 2 to 3 mm. The diameter of the isthmus is about one-half that of the ampulla. The entire length of a Fallopian tube varies from about 15 to 30 cm in the various farm mammals. It is obviously somewhat shorter in the smaller mammals.

The Fallopian tubes serve several purposes. An obvious function is to carry the ovum from the ovary to the uterus. It also transports sperm cells from the uterus toward the infundibulum after copulation. The upper end of the Fallopian tubes is where the ovum becomes fertilized. The fluids secreted in the Fallopian tubes provide a suitable environment for the fertilized ovum until it reaches the uterus. After fertilization, the zygote will remain in the Fallopian tubes from two to four days until it finally passes into the uterus.

The Uterus

The mammalian uterus consists of the body, two uterine horns, and the cervix. The shape and relative proportions of the various parts vary from one species to another. Figure 8-2 illustrates some of these differences. In the embryonic stages of the development of the reproductive system, the female tract consists of two tubes that come together to form a single tube. The appearance is that of a Y-shaped tubular structure. The two upper parts of the embryonic tract develop into the Fallopian tubes and the two uterine horns. The lower or single part of the tract develops into the vagina and body of the uterus. The junction where the two tubes become one always occurs within the uterus. The exact point at which the junction occurs determines the relative sizes of the uterine horns and the body of the uterus. The uteri of litter-bearing animals generally have very prominent horns and relatively small bodies. The horns of the uterus in the cow and the ewe are relatively less prominent than those of the sow and the bitch. However, in each of the four

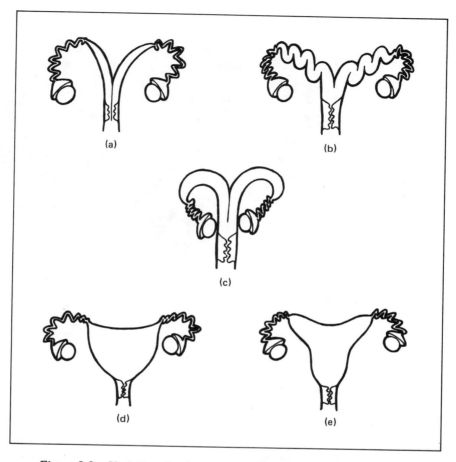

Figure 8-2 Variations in the structure of uteri of domestic animals.
(a) Duplex uterus. (b) A sow's bicornuate uterus. (c) A cow's bicornuate uterus. (d) Simple uterus. (e) Bipartite uterus. [H. Joe
Bearden and John W. Fuquay, *Applied Animal Reproduction.*
(Reston, Virginia: Reston Publishing Company, Inc., 1980) p. 14]

species, pregnancy almost always occurs in the uterine horns (see
Figure 8-3). In the mare, the body of the uterus is more prominent
than the horns and pregnancy will more often occur in the body. In
the human uterus, there are no uterine horns. In all species the uterus
is supported in the body cavity by that part of the broad ligament
called the mesometrium.

     The primary function of the uterus is that of providing a place

Figure 8-3   Reproductive tract of a sow pregnant for 28 days.
(Courtesy University of Missouri)

for the developing embryo or fetus until it is ready to live outside
the body of the dam. The uterus also serves as the passageway for
sperm on their way from the vagina to the Fallopian tubes. Contractile activities of the uterus help in this transporting process.

The uterine wall consists of three kinds of tissue: the outer
layer, known as the serosa; the middle layer, called the myometrium;
and the inner layer, called the endometrium. The myometrium consists of smooth muscle. The activity of the uterus that occurs during
the heat period and at the time of parturition depends upon contractions of this muscle layer.

The inner layer, or endometrium, is the part of the uterus that
is in contact with the fetal membranes during pregnancy. It is well
supplied with blood vessels and changes in thickness and vascularity
with changes in the estrous cycle and during pregnancy. The inside
surface of the endometrium of cows, ewes, and other ruminants
contains several rows of small structures called caruncles. During
pregnancy, these caruncles form the uterine part of the attachment
to the fetal membranes. The endometrium of the mare and sow has
no caruncles. The fetal membranes in these species form their attach-

ment with the entire surface of the endometrium rather than with specialized areas as in ruminants.

The cervix may be referred to as the mouth of the uterus. It is a thick-walled structure composed of several folds and interlocking annular rings. It is sealed tightly during pregnancy to keep any foreign material out of the uterus. It has a tremendous capability to stretch when under the influence of the proper balance of hormones such as at the time of parturition. Along with the vagina, it serves as a part of the birth canal.

The Vagina and Vulva

The vagina is the female organ of copulation, or coitus. It is inside the vagina where semen is usually deposited by the male of the species. The boar is one of the few animals that at least partially penetrates the cervix during copulation. The urethra opens into the floor of the vagina about two-thirds of three-fourths of the way from the cervix to the vulva. In females, the urethra is not a part of the reproductive tract as it is in males.

The entrance to the vagina is protected by two pairs of door-like structures. The outermost pair of folds are called the labia majora. These are the visible parts of the vulva. Just inside the outer folds of tissue are the second pair, called the labia minora. Located just inside the vulva on the floor of the vagina is the clitoris. The clitoris will sometimes be embedded in the floor of the vagina and may be difficult to locate. It is made up of erectile tissue similar to that of the male penis. In fact, the clitoris has the same embryonic origin as the penis. It is well supplied with nerves and plays a role in the sexual stimulation of the female during copulation.

STRUCTURE AND FUNCTION OF THE OVARY

The mammalian ovary is attached to that part of the broad ligament called the mesovarium. The ovary is attached by means of a stalk called the hilus. The ovary of a cow is about the size of the first segment of a human thumb. The ovary of a sow or a ewe is about two-thirds the size of that of a cow. In a mare the ovary is about twice as large as that of a cow.

The ovary has an inner part called the medulla and an outer layer called the cortex. The medulla consists of connective tissue, blood vessels, and nerves, In the cortex are found the follicles in various stages of development. The corpora lutea would also be found in the cortex. Around the outside of the ovary is the germinal epithelium. This is the part of the ovary from which the primary follicles began their development before the female was born. The follicles and the corpora lutea are the source of the female sex hormones. The female gamete, the ovum, develops in the follicle.

Follicular Development and Ovulation

By the time of birth of a female mammal, hundreds of primary follicles are present beneath the outer layers of tissue surrounding the ovary. It is generally thought that no new primary follicles will develop in the ovary after birth. A primary follicle consists of a germ cell surrounded by a single layer of cells. Figure 8-4 shows the developing follicle in several of its various stages. The germ cell, which will eventually develop into the ovum, is referred to as an ovigonium. Ovigonium is the Latin term for this cell. It is much easier to pronounce than the Greek term oögonium used by many authorities. However, either term is appropriate.

The follicles do not develop further than the primary stage until after puberty or sexual maturity. During each cycle thereafter, some of the primary follicles begin development under the influence of hormones from the pituitary gland. In mares, cows, women, and other mammals that usually have only one offspring per pregnancy, only one follicle generally reaches the mature stage and ovulates. In ewes, one to three mature follicles may develop. In litter-bearing animals, 10 to 25 mature follicles may be produced. As a primary follicle proceeds to develop, the outer layer of cells around the ovigonium begins to grow. The follicle increases in diameter as several layers of cells develop around the future ovum. At this stage of development, it is called a secondary follicle. As the follicle continues to develop and enlarge, a cavity, or antrum forms inside. The antrum will be filled with a fluid called follicular fluid. The

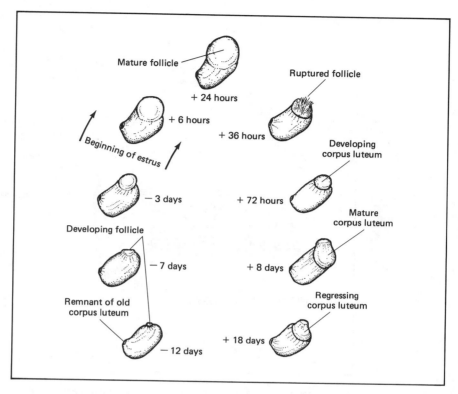

Figure 8-4 Ovary of a cow showing the progressive development of a follicle and corpus luteum. Time begins with estrus at 0 hour.

follicle is now called a tertiary follicle. At this stage the follicle forms a blisterlike protrusion on the surface of the ovary.

The mature follicle is shown in Figure 8-5. The size of the mature follicle varies with species, but may be almost as large as the ovary itself. The mature follicle is also called a Graafian follicle. The protruding follicle is covered by the germinal epithelium of the ovary. The wall of the follicle itself is a two-layered membrane called the theca membrane. The internal layer of the theca secretes the female sex hormone, estrogen. The estrogen is absorbed by the blood supply to the ovary and carried throughout the body of the animal. The cavity of the follicle is lined with a layer of granular cells called the granulosa membrane. Some of the granular cells inside the follicle form a pedestal upon which is found the developing ovum.

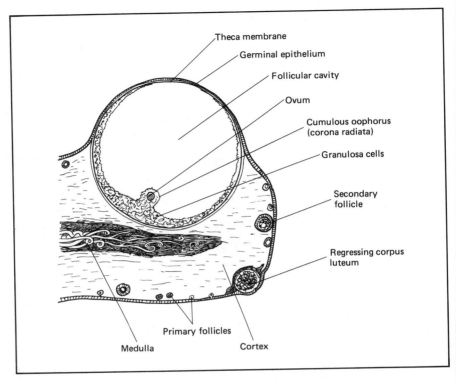

Figure 8-5   Section of an ovary showing a mature follicle.

These cells also surround the outer wall of the ovum and are referred to collectively as the cumulous oöphorus. This cumulous mass will remain with the ovum until after ovulation.

At the proper stage of the reproductive cycle, the follicle ruptures and the ovum is released into the infundibulum of the Fallopian tube. This is called ovulation. In most species there will be a small amount of bleeding in the follicular cavity following ovulation. A small clot usually forms but is absorbed within a few days. The clot is referred to as the corpus hemorrhagicum or "body of blood." Several other follicles which may have been developing will not become fully developed and will regress and be absorbed by the ovary. Figure 8-6 is a photograph of the ovary of a sow showing a large follicle and several corpora lutea.

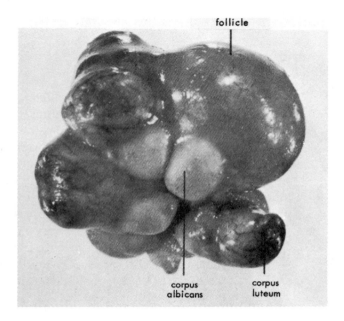

follicle

corpus
albicans

corpus
luteum

Figure 8-6   A sow ovary showing follicles and corpora lutea.
(Courtesy University of Missouri)

Development of the Corpus Luteum

Within the crater where the mature follicle ruptured and ovulated, a
new kind of tissue begins to develop. Because of the blood clot that
is present, for a few days it is referred to as the corpus hemorrhagicum. As the blood and old follicular tissue is reabsorbed, lutein cells
begin to develop. In cows the corpus luteum will be fully developed
about a week after ovulation. The term corpus luteum literally means
"yellow body," but in some animals it may be pink. The corpus
luteum is the source of the second female sex hormone, progesterone.

If pregnancy does not occur, the corpus luteum will remain
functional for about eight to ten days in the cow and then begin to
regress. As the corpus luteum regresses, a new follicle will start to
develop to begin another cycle. If pregnancy does occur, the corpus
luteum remains functional throughout most of pregnancy. After the
corpus luteum regresses it forms a small white scar on the ovary
called the corpus albicans (see Fig. 8-6).

Ovigenesis (Oögenesis)

The ovum begins as a germ cell called the ovigonium inside the primary follicle. As the follicle develops, the process of ovigenesis proceeds inside it. In a mature female, the first stage in the process of ovigenesis is the development of the primary ovicyte. This involves an increase in the amount of cytoplasm and the formation of a relatively thick membrane around the ovicyte called the zona pellucida. Much of the increase in the volume of cytoplasm comes from the production of yolk granules or deutoplasm. The purpose of the yolk is to provide nutrients for the embryo until it reaches the uterus. The cytoplasm of the developing ovicyte is called vitellus.

The diameter of the primary ovicyte varies with species from about 120 to 200 μ. The zona pellucida around the cell is about 10 μ thick. This thick membrane protects the ovicyte as it develops and for several days after it becomes fertilized. It also provides a strong container to hold the cell and developing embryo intact. A cell as large as the ovicyte or zygote would very likely rupture if all it had to hold it together was a cell membrane of the usual thickness of a typical body cell. A typical cell may be about 10 μ in diameter with a cell membrane that is 100 Å or 0.01 μ thick. Inside the zona pellucida there is a second membrane called the vitelline membrane which is about the same thickness of a typical cell membrane. By the time the primary ovicyte is fully formed, the follicle is about in the tertiary stage.

The first division of ovigenesis produces a secondary ovicyte that retains all of the vitellus or cytoplasm of the primary ovicyte. One-half the contents of the nucleus is expelled into the space between the vitelline membrane and the zona pellucida. This space is called the perivitelline space. The extruded nucleus is called the first polar body. The only purpose of the first polar body is to be a carrier of one-half the chromosomes produced by the first division of meiosis. The polar body may or may not undergo a second division. Either way, it will be reabsorbed by the zygote later.

By the time the secondary ovicyte is ready for the second division, ovulation will have occurred in most mammals. The second division only occurs after a sperm cell penetrates the vitelline membrane. If fertilization fails to occur, the secondary ovicyte is carried part of the way down the Fallopian tube but dies and is reabsorbed before it can reach the uterus. If fertilization does occur, the second division follows, resulting in the extrusion of the second polar body into the perivitelline space. If the first polar body should also divide, it is possible for the zygote to have three polar bodies trapped beneath the zona pellucida for a period of time. In some animals, such as the dog and the mare, the cell released from the mature follicle at the time of ovulation is actually the primary ovicyte.

The Ovum

When the ovum is released from the follicle it is covered with several layers of granular cells from inside the follicle called the cumulus oöphorus. The cells adhere to the zona pellucida in such a way as to give it a sunburst appearance. Because of this appearance, the term corona radiata is used to describe the material surrounding the cell. The cell mass usually adheres to the zona for only a few hours before it is worn away by the action of the Fallopian tubes.

The term ovum as it is normally used refers to the cell that is released from the mature follicle at ovulation. Technically, a true ovum never really exists. The cell that leaves the follicle is a secondary ovicyte in most animals and a primary ovicyte in some. In order for a true ovum to exist, the second division must have occurred. By the time that happens, a sperm cell will have already entered the so-called ovum which means it is not an ovum, but a zygote. Sometimes the term ovum is even used when referring to a zygote or a three- or four-day-old embryo. As far as their external appearance goes, the primary ovicyte, the secondary ovicyte, the zygote, and the embryo which is several days old generally look alike. The zona pellucida is present throughout this whole period of about a week. The cell may have no polar bodies or it may have one, two, or three polar bodies. It may even be an embryo with eight or sixteen or more cells. As it is commonly used, then, the term ovum refers to the ovicyte that is released at ovulation and may continue to be used to refer to that structure for several days thereafter, whether it be fertilized or not. An unfertilized ovum and a developing embryo are shown in Figure 8-7.

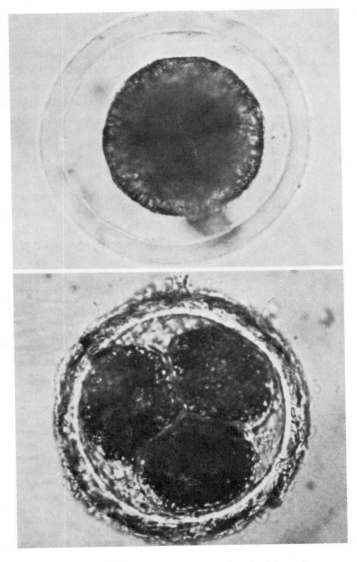

Figure 8-7   Unfertilized ovum and developing embryo.
(Courtesy University of Missouri)

REPRODUCTIVE CYCLES

Comparisons of Reproductive Characteristics in Some Mammals

Puberty is that point in the life of an animal when it is able to produce viable gametes and carry out the reproductive function. In nonprimate female mammals it is marked by their first estrus, or heat period. The female experiences her first estrous cycle at puberty. It is associated with an enlargement of the reproductive tract and development of one or more mature follicles. The age of an animal at puberty varies within species. Estimates of this trait for selected species are included in Table 8-1.

The reproductive cycle in nonprimate mammals is called the estrous cycle. It is measured from the beginning of one estrus to the beginning of the next estrus in a nonpregnant animal. Estrus is that period during the cycle when the female is sexually receptive to the male; in other words, she is willing to perform coitus or copulation. The length of the estrous cycles of several mammals is shown in Table 8-1. In cows, sows, and mares it will average about 21 days. In ewes it is several days shorter.

The phase of the cycle just before estrus is called proestrus. It is during this phase that the follicles are developing. Following estrus is the phase called metestrus. During this phase the corpus luteum develops and the uterus is prepared for pregnancy. If pregnancy does not occur, the reproductive tract returns to a less active state to await the beginning of the next cycle. This last phase is sometimes called diestrus. Some sources define metestrus as only that period of time following estrus during which the corpus luteum is developing, and diestrus as the remaining part of the cycle that extends until the next proestrus. Other sources do not refer to a proestrus of a metestrus but use the term diestrus to refer to the entire time between two heat periods.

Most farm mammals are referred to as continuous breeders. This means that unless they become pregnant, they will exhibit estrous cycles the year round. And because they have many cycles each year they are also said to be polyestrous. Ewes are polyestrous, but are seasonal breeders. During the season of the year when the ewe is not cycling she is said to be in anestrus. Dogs are also seasonal breeders,

TABLE 8-1.   CHARACTERISTICS
OF THE REPRODUCTIVE CYCLES
OF SELECTED MAMMALS

Animal	Age at Puberty (months)	Length of Estrous Cycle (days)	Duration of Estrus	Time of Ovulation	Number of Ova Shed
Cow	4-14	21 (18-24)	18 hours (3-28)	4-16 hours after estrus	1
Ewe	7-12 (first fall)	17 (14-20)	30 hours (24-36)	12-24 hours before end of estrus	1-3
Sow	5-8	21 (18-24)	2 days (1-5)	16-48 hours after start of estrus	10-20
Mare	15-24	21 (15-24)	5 days (3-9)	1-2 days before end of estrus	1
Goat	7-12	18 (16-22)	45 hours (30-60)	second day of estrus	2-3
Dog	6-24	—[a]	9 days (5-19)	1-2 days after start of estrus	8-10
Rabbit	—	15 (14-16)	30 days	10 hours past coitum	10
Rat	—	5 (4-6)	9-20 hours	8-11 hours after start of estrus	10

[a]Two breeding seasons yearly with one cycle and one estrus during each.

but monestrous. This means that there is only one estrous cycle during their breeding season. Dogs have two breeding seasons yearly with one cycle during each. Most wild animals are seasonal breeders. Their breeding season occurs at a time such that the young will be born during the early spring months. In squirrels and rabbits, and other animals with short gestation periods, the breeding season may occur in the winter or very early in the spring. With deer and other animals with longer gestation periods, the breeding season will be during the fall and late summer months.

The reproductive cycles of primates (chimpanzees, gorillas, monkeys, and humans) are very similar to those of other mammals with at least two exceptions. Primates do not exhibit estrus, so are usually sexually receptive at all times. They also exhibit menstruation whereas other mammals do not. The reproductive cycle is thus called a menstrual cycle and is measured from the beginning of one menstrual period to the beginning of the next. If pregnancy does not occur in a primate, there is a loss of blood and tissue from the inner lining of the uterus. This expulsion of blood and tissue is called menstruation.

Reproductive Cycle in Nonpregnant Mammals

The estrous and menstrual cycles are under the control of hormones produced by the anterior lobe of the pituitary gland. These are the same hormones that control the activities of the testes in the male. Because of their effects on the gonads, they are called gonadotropic hormones. Follicle-stimulating hormone, or FSH, stimulates the development of the follicle and the growth of the ovicyte inside the follicle. The theca cells, or theca membranes, of the follicle are also stimulated by FSH to secrete estrogen. As the follicle increases in size it produces more and more estrogen. When the estrogen level in the blood reaches a certain point it results in estrus. Estrus coincides approximately with ovulation. This is to insure that the sperm cells will be deposited in the female at the time when the ova or ovum will be available for fertilization. When the follicle ruptures at ovulation, the estrogen levels decline substantially and the female soon ceases to show signs of estrus. Cows are an exception in that their estrus normally terminates before ovulation.

The duration of estrus and the times of ovulation for several species are presented in Table 8-1. Estrogen stimulates muscular activity of the reproductive tract to aid in the movement of the sperm from the vagina, through the uterus, and up the Fallopian tubes. Estrogen also stimulates the development of secondary sex characteristics, or the feminine characteristics, of the female. Over a period of time, estrogen stimulates increased growth and development of the reproductive tract and the mammary glands.

When the follicle has developed to the point that it is ready to ovulate, LH, or luteinizing hormone, is released from the anterior pituitary. The mechanism of LH release is not fully understood but it may be that the increase in the estrogen level in the blood plays a part. Luteinizing hormone causes the wall of the follicle to rupture and release the ovum. This is another way of saying that LH brings about ovulation. This hormone also stimulates the development of the corpus luteum and the secretion of progesterone by the corpus luteum.

Progesterone is sometimes called the hormone of pregnancy. It brings about the necessary changes in the uterus in preparation for the arrival of the embryo from the Fallopian tube and the eventual implantation of that embryo in the uterus. The preparation of the uterus for pregnancy includes a thickening of the endometrium, an increase in the blood supply to the uterus, and an increase in the secretory activity of the endometrium. In addition, toward the end of pregnancy, progesterone plays a role in the development of the mammary glands in preparation for lactation.

In nonpregnant ewes, the corpus luteum begins to regress in about 12 days after the beginning of estrus. This regression naturally results in a decrease in the progesterone level in the blood. This in turn tends to cause the anterior pituitary to release more FSH to stimulate the development of one or several new follicles. The changes that occurred in the uterus in anticipation of pregnancy are reversed during this period of the cycle. The thickness of the endometrium of nonprimates is reduced and blood supply diminishes. In primates, this material is not reabsorbed but released and expelled, causing menstruation.

PREGNANCY AND PARTURITION

The period of pregnancy, or gestation, begins with the fertilization of the ovum by the sperm and ends with the process of parturition. During pregnancy, the fetus obtains oxygen and other vital nutrients from the blood of the dam in exchange for carbon dioxide and other waste products which are absorbed by the mother's blood. This phenomenon is unique to mammals. Among the domestic and laboratory animals, the length of gestation varies from 3 weeks in rats to 11 months in mares, as shown in Table 8-2.

TABLE 8-2. THE AVERAGE LENGTH
OF GESTATION IN DOMESTIC MAMMALS

Mammal	Length of Gestation
Rat	21 days (3 weeks)
Rabbit	31 days (1 month)
Dog and cat	63 days (2 months and 2 days)
Sow	114 days (3 months, 3 weeks, and 3 days)
Ewe	147 days (4 months and 27 days)
Cow	282 days (9 months and 9 days)
Mare	335 days (11 months)

Fertilization to Implantation

Fertilization occurs in the upper one-third of the Fallopian tube. Within 15 minutes after copulation or insemination, sperm cells can be found in the upper end of the Fallopian tube. Contractions and other activities of the uterus and Fallopian tubes probably play the major role in the movement of the sperm through the tract. In most animals, ovulation occurs after the beginning of estrus. If the opportunity for mating is available, this means that insemination will very likely occur before ovulation. This is one reason why fertilization occurs in the upper one-third of the Fallopian tube. Another reason why fertilization occurs there relates to the life-span of the ovum and the amount of time required for it to move down the Fallopian tube. The fertilizable life of the ovum extends for about one day beyond ovulation. If it is not fertilized within this period it will die and be reabsorbed by the reproductive tract. If it remains alive, the amount of time required for the ovum to pass through the Fallopian tube is about three days. If the rate of travel of the ovum is about the same throughout the length of the Fallopian tube, it will travel about one-third of the length of the tube in one day. This means that the cell must be fertilized while it is in the upper one-third of the tube or it will not live to be fertilized.

The process of fertilization involves the passage of a sperm cell through the zona pellucida and vitelline membrane and, finally, its union with the nucleus of the ovum. At this point the fertilized cell becomes a zygote. Several thousand sperm cells may be available to fertilize the ovum but only one ultimately accomplishes the task. In some species, such as sheep and dogs, after one sperm succeeds in penetrating the zona pellucida no other sperm can enter. The ability of the zona to resist the penetration of additional sperm is called the zona reaction. In other species, a zona reaction exists in varying degrees. Some are quite effective in keeping additional sperm cells out while others are not. Rabbits show no zona reaction. In those species where additional sperm do manage to penetrate the zona pellucida, there is a second preventive reaction known as the vitelline block. Once a single sperm cell has penetrated the vitelline membrane, additional sperm cells cannot enter.

These measures to prevent the entrance of extra sperm cells are necessary to avoid the condition called polyspermy. Polyspermy would result in too many sets of chromosomes in the zygote should two or more sperm cells actually fertilize the ovum. A normal zygote is diploid. If an ovum should become fertilized by two sperm cells a condition called triploidy would exist. This happens on rare occasions, but when it does the embryo usually dies early in its development.

After the zygote is formed, it begins to divide mitotically. The division occurs in such a way that the total size of the embryo does not change. Every division results in a decrease in the size of the daughter cells. This phase of embryonic development is called cleavage.

By the time the embryo reaches the uterus, it has developed to about the 16-cell stage. Soon after this the cells begin to differentiate and the various body tissues begin to develop. By day 21 after fertilization in cattle embryos the heart begins to beat. By day 28, all four limb buds are visible. The study of the development of the embryo is called embryology.

Before implantation takes place, the embryo is nourished while inside the uterus by secretions of the uterus called uterine milk. Before the embryo reached the uterus, it obtained its supply of nutrients from the yolk granules, or deutoplasm, that originated from the cytoplasm of the original ovicyte.

Three sets of membranes develop around the embryo as it prepares for attachment to the wall of the uterus. A diagram of these

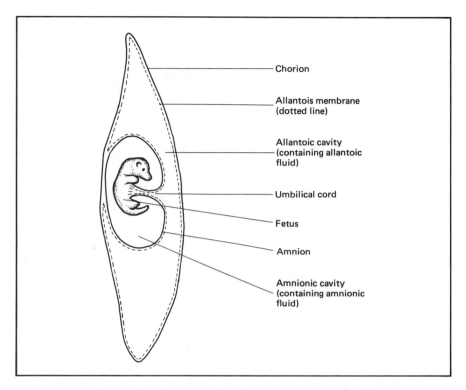

Figure 8-8   Diagram of the fetal membranes of a pig.

membranes is shown in Figure 8-8. The outermost of these membranes is called the chorion. This is the layer of the embryo that comes in contact with the endometrium of the uterus. The embryo itself is surrounded by a fluid-filled sac called the amnion. The fluid in the amnion, called amnionic fluid, protects the embryo from shock. The amnion is connected to the embryo at the navel by a stalk called the umbilical cord.

The allantois makes up the third set of membranes. It is an extension of the urinary system that grows out through the umbilical cord and lies between the amnion and chorion. As pregnancy progresses, the allantoic sac increases in size and occupies more of the space between the other two membranes. The allantoic sac also contains fluid that originates from the kidney of the embryo or fetus. The term fetus is used to refer to the embryo after the body

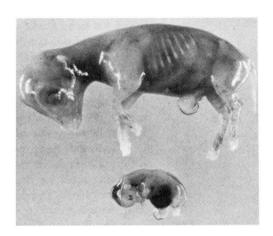

Figure 8-9   A fetal calf at 75 days after fertilization compared to a 42-day embryo. [H. Joe Bearden and John W. Fuquay, *Applied Animal Reproduction.* (Reston, Virginia: Reston Publishing Company, Inc., 1980) p. 93]

organs have formed. Figure 8-9 shows calf embryos at 42 and 75 days after fertilization.

In species—such as cattle, horses, and humans—that usually produce one offspring at a time, occasionally twins are born. Two types of twinning exist: monozygotic and dizygotic. Dizygotic twins result from the fertilization of two ova by two separate sperm cells. This is the most common type of twinning. From about 2 to less than 1 percent of all births in cattle are twins. Variations exist among breeds and types of cattle. The incidence of twinning in humans is about the same as it is in cattle. Dizygotic twins are no more alike genetically than any two sibs (sisters and/or brothers) born at different times. They may, however, be more alike for certain factors controlled by the environment since they are the same age, have shared the same uterine environment, and are more likely to share similar environments after birth.

Monozygotic twins, also called identical twins, develop from one embryo. Because they each have the same genetic makeup, whatever differences they may show are due to environmental factors. The incidence of this type of twinning is much less frequent than dizygotic twinning. Possibly only 1 to 5 percent of all twin births are monozygotic.

Implantation to Parturition

At about two weeks to two months after fertilization in farm animals an attachment forms between the chorion of the embryo and the endometrium of the uterus. The chorion develops hairlike projections called villi that extend into the endometrium. This process of attachment requires about one to three weeks to complete and is called implantation. Collectively, the area of attachment is referred to as the placenta. Blood vessels from the fetus extend through the umbilical cord to the surface of the chorion to form capillaries in the villi. The blood supplies of the fetus and dam do not mix, but nutrients and oxygen pass from the blood of the dam across the membranes of the placenta into the blood of the fetus. Waste products and carbon dioxide pass from the fetus to the dam in a similar manner.

The types of placenta vary among mammals. The sow and mare have a type called the diffuse placenta. The entire surface of the chorion develops villi and makes contact with the endometrium. In dogs, the area of attachment is limited to a band or zone that extends around the chorion. This type is called the zonary placenta. Rabbits and women possess a discoid placenta. The area of attachment is limited to a circular patch or disc at one end of the chorion. Implantation in women differs from that in farm animals in that the human embryo literally burrows its way into the endometrium. The development of the fetus in women does not actually occur in the lumen of the uterus as in farm mammals.

Ruminants have the cotyledonary type of placenta. Specialized areas on the surface of the chorion called cotyledons develop and form an attachment with the caruncles that exist permanently on the wall of the uterus. From 70 to 120 caruncles exist in the uterus of a cow. A single cotyledon and its adjacent caruncle is called a placentome. All of the placentomes are referred to collectively as the placenta.

The placental tissues of some animals produce certain hormones. Two examples are of particular interest. The endometrium of the pregnant mare secretes an FSH-like hormone between about days 40 to 160 in the pregnancy. The hormone can be detected in the blood. It is called pregnant mare's serum or PMS and can be used as a source of FSH-like gonadotropin. The chorion of the pregnant

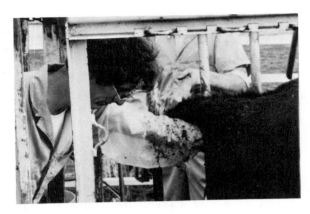

Figure 8-10   Rectal palpation of a cow.

woman secretes an LH-like gonadotropin through most of pregnancy beginning during the second month. It is called human chorionic gonadotropin or HCG.

The presence of hormones in the blood or urine makes it possible to diagnose pregnancy in mares and women. The injection of mare's serum into immature laboratory animals will stimulate follicular development if the mare is pregnant. After 120 days of pregnancy, a chemical test will reveal the presence of estrogens in the urine of pregnant mares. No reliable hormone tests are available for cows, sows, or ewes. Experienced technicians can palpate the uterus and ovaries of cows and mares through the rectum and not only determine whether they are pregnant, but give reliable estimates of the age of the fetus (see Fig. 8-10). Ultrasonic means can be used to detect pregnancy in sows and ewes as shown in Figure 8-11.

Parturition

The act of giving birth, parturition, is usually referred to by terms appropriate to the species such as calving, foaling, and lambing. In swine it is called farrowing. In dogs it is called whelping. Pregnancy, or gestation, terminates with parturition. The time of parturition with respect to fertilization varies some within species but, when compared to the length of gestation, in most species the variation is relatively small.

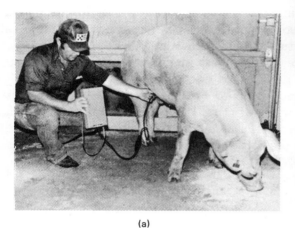

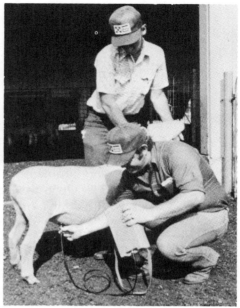

(a)   (b)

Figure 8-11   Checking (a) a sow and (b) a ewe for pregnancy using
ultrasonic equipment. [H. Joe Bearden and John W. Fuquay, *Applied
Animal Reproduction.* (Reston, Virginia: Reston Publishing Company, Inc., 1980) pp. 250 and 254]

Parturition is brought on by a number of factors, many of
which are not fully understood. Size and age of the fetus may play
a role. Twins in cattle and sheep are usually born a few days sooner
than singles. A larger calf may be born before a smaller one. However, bull calves are usually larger than heifers but are carried one
or two days longer. As the uterine contents increase in size, they are
throught to create pressures on the cervix and uterine wall which are
a factor in bringing about parturition.

Certain hormones are known to play a part in parturition.
Oxytocin, a hormone released from the posterior pituitary, causes
contraction of the myometrium of the uterus. Oxytocin in mammals
is very similar to vasotocin in birds which aids in the egg-laying
process. Progesterone is necessary in most species for the maintenance of pregnancy. As the time for parturition approaches, the
progesterone level in the blood usually declines. The question arises
as to whether this decline has something to do with initiating parturition or whether the other factors causing parturition cause the

213

decline in progesterone level. In some species a hormone called relaxin is produced by the ovary. It stimulates the relaxation of the cervix and pelvis to allow the fetus to be expelled more easily.

Occasionally, in mares and cows, because of the size or position of the fetus, parturition may be difficult. Young females and particularly small females bred to large males may have problems. The ideal presentation of the fetus at parturition is for the head and front feet to exit first. If both rear feet appear first, a normal birth can occur with no problems if the process does not take longer than a few minutes. Once the umbilical cord breaks, the young mammal must start breathing or it will suffocate. Any presentation of the fetus other than the two just described is considered abnormal and will result in difficulty in parturition. Difficult parturition is sometimes called dystocia.

Estrus Synchronization, Superovulation, and Embryo Transfer

Synchronization of estrus is a process whereby an attempt is made to adjust the estrous cycle of all the females in a herd so that they will all exhibit estrus at about the same time. If this can be done, it would allow all of them to be bred at the same time and thus have their offspring born at about the same time. The result would be greater uniformity among the offspring and a possible saving of labor and time at calving or farrowing time.

Synchronization of estrus would also allow more efficient use of artificial insemination. In beef cattle, for instance, it is often difficult to observe cows in estrus, which in turn makes it difficult to use artificial insemination effectively. If cycles could be adjusted so that all cows were to come into estrus at some prearranged time, they could all be inseminated within a one- or two-day period.

One method that is being used to synchronize estrus in cows and mares involves the injection of prostaglandin. This is a substance that is normally produced by the body but can also be synthesized commercially. Proper use of prostaglandins causes the corpora lutea to regress and, in effect, causes the estrous cycle to begin all over again. The ultimate effect is that all treated females usually exhibit estrus at about the same time.

Figure 8-12   Flushing a cow for embryos.

Superovulation is a process whereby females are injected with gonadotropins to cause them to produce more ova than they normally would naturally. The injection of proper dosages of FSH during proestrus can bring about the development of numerous follicles. At the proper stage of development, injections of LH will cause ovulation of these follicles.

Synchronization of estrus and superovulation are now being used in conjunction with another process called embryo transfer. In cattle, for instance, one or more superior cows and a number of ordinary healthy recipient cows are synchronized. The superior cows are superovulated and bred to superior bulls by artificial insemination. In about a week after insemination, a number of embryos are removed from the superior donor cows by flushing the uterus with a special salt solution. Over 20 live and healthy embryos can sometimes be obtained, but the average number obtained is normally in the range of 8 to 10. By a procedure similar to artificial insemination, the live embryos are transferred to the recipient cows that are in the same stage of the estrous cycle as the donor cows. The embryos usually develop through a normal pregnancy in the recipient cows. After a period of time, the process of superovulation and embryo collection can be carried out again with the superior cow. In this way, one superior cow might be able to produce 40 or 50 or more calves in one year and have them borne and raised by less superior cows. Figure 8-12 shows part of the process of flushing a cow for embryos.

STUDY QUESTIONS AND EXERCISES

1. Define each of the following terms:

allantois

amnion

ampulla

antrum

anestrus

broad ligament

caruncle

cervix

chorion

cleavage

clitoris

conception

copulation

corona radiata

corpus albicans

corpus hemorrhagicum

corpus luteum

cotyledon

cumulous oöphorus

deutoplasm

diestrus

dizygotic twins

dystocia

embryo

embryo transfer

endometrium

estrogen

estrous cycle

estrus

Fallopian tube

farrow

fertilization

fetus

follicle

FSH

gestation

HCG

heat

identical twins

implantation

infundibulum

isthmus

LH

menstrual cycle

menstruation

metestrus

monoestrous

monozygotic twins

myometrium

ovary

ovicyte (oöcyte)

oviduct

ovigenesis (oögenesis)

ovigonium (oögonium)

ovulation

ovum

oxytocin

parturition

placenta

placentome

PMS

polar body

polyestrous

polyspermy

pregnancy

proestrus

progesterone

prostaglandin

puberty

relaxin

serosa

sib

superovulation

theca

triploidy

umbilical cord

uterus

vagina

villi

vitelline membrane

vitellus

vulva

yolk

zona pellucida

zygote

2. Describe the general organization of the reproductive system of a female mammal.

3. Describe the functions of the Fallopian tubes including the various parts.

4. Compare the structures of uteri in several domestic animals.

5. Compare the functions of the reproductive tract of the chicken to that of the sow.

6. Describe the three regions of the broad ligament, giving their names and functions.

7. Describe the process of ovigenesis.

8. Describe the stages in the development of an ovarian follicle.

9. Compare the size of an ovum and its membranes to that of a typical body cell.

10. Describe the various events that occur during the estrous cycle of a sow.

11. Explain why fertilization in farm mammals occurs in the ampulla or infundibulum.

12. Describe and explain the function of the zona reaction and the vitelline block.

13. Compare monozygotic and dizygotic twinning.

14. Compare the various sources of nutrition for the embryo and fetus from the time of fertilization until after implantation.

15. Describe and compare the several kinds of placentae found in domestic animals.

16. What causes parturition?

17. Describe the process of estrus synchronization.

18. Discuss the advantages of embryo transplant.

9

The Male
Reproductive System

Individual animals originate from a single cell called a zygote. A zygote is formed by the union of two different sex cells, one of which is produced by a male parent, the other by a female parent. The reproductive systems of animals deal with the formation and transmission of these sex cells and ultimately their union to form the zygote. In birds and mammals the male sex cell, or gamete, must be transferred from the reproductive system of the male to that of the female. The male system is designed to carry out that process.

For the purposes of discussion and explanation, the male reproductive system can be organized into four divisions: the gonads, the reproductive tract, the accessory glands, and the penis or copulatory organ. The gonads of the male are called testes or testicles. Their function is to produce the male gametes or sperm cells, and the male sex hormone, testosterone. The reproductive tract carries the gametes from the testes to the penis to be transferred to the female tract. The sex act, or sexual intercourse, is called copulation. The accessory glands function in producing fluids that make up the major part of the seminal plasma. Diagrams of the reproductive systems of the bull and stallion are shown in Figures 9-1 and 9-2.

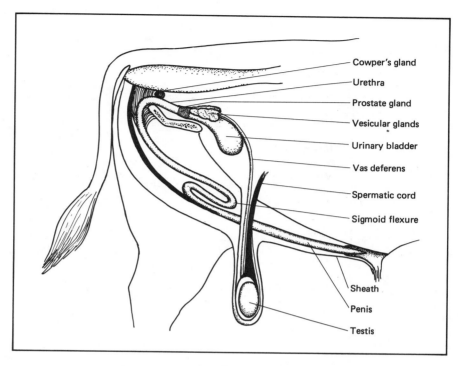

Figure 9-1   Reproductive organs of a bull.

THE TESTES AND SCROTUM

The Scrotum and Spermatic Cord

In mammals, a pair of testes are located inside a special pouch called the scrotum which is suspended outside the body cavity. In some species such as mice, rats, and rabbits, it may be referred to as the inguinal pouch. The wall of the scrotum consists basically of three types of tissue. The outermost layer of tissue is the skin. The innermost layer of the scrotum is called the tunica vaginalis. The term tunic means coat or covering. This layer of tissue is actually continuous with the abdominal lining that extends through the inguinal canal and into the scrotum. The term vaginalis suggests that the abdominal lining, called the peritoneum, evaginated from

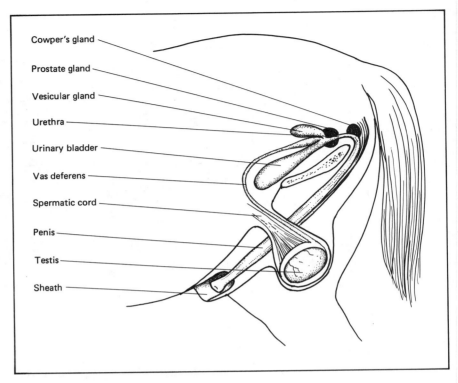

Cowper's gland

Prostate gland

Vesicular gland

Urethra

Urinary bladder

Vas deferens

Spermatic cord

Penis

Testis

Sheath

Figure 9-2   Reproductive organs of a stallion.

the abdominal cavity to form a pocket inside the scrotum as the testes passed from the abdominal cavity into the scrotum.

Between the tunica vaginalis and the skin lies a layer of flat muscular tissue called the tunica dartos. This muscle functions in the heat regulatory mechanism of the testes. In most mammals, the temperature of the testes must be maintained within rather narrow limits, at a temperature several degrees below that of the body, in order for spermatogenesis to take place. The temperature is regulated by the contraction and relaxation of the tunica dartos muscle which causes the testes to be raised or lowered with respect to the body. In very cold weather, the scrotum of a bull may be pulled very close to the body wall and take on a highly wrinkled appearance. In very hot weather, the scrotum becomes quite pendulous and will be suspended some distance from the body wall. In most mammals the scrotum is not so pendulous as in the ram and bull. Birds do not possess a scrotum; their testes remain inside the body cavity.

In most mammals the testes descend from the abdominal cavity into the scrotum at or near the time of birth. Failure of the testes to descend is referred to as cryptorchidism. An animal with both testes remaining in the abdominal cavity is known as a bilateral cryptorchid. Because of the exposure of the testes to the higher temperature inside the body cavity, spermatogenesis cannot take place in a bilateral cryptorchid. However, the production of the male sex hormone is not altered by the increase in the temperature of the testes, so such an animal will maintain a normal sex drive.

If only one testis fails to descend and the second is carried in the scrotum, the animal is called a unilateral cryptorchid or monorchid, and is fertile. It is not wise to use monorchids for breeding purposes, however, because the tendency toward cryptorchism is heritable. In some breeds of livestock the cryptorchid condition can result in disqualification for registration. The testes may fail to descend because the opening in the abdominal wall through which they must pass is too small to accommodate them. This opening is called the inguinal canal. In most species of domestic animals, once the testes have descended, the inguinal canal becomes much smaller, preventing the testes from passing back into the abdominal cavity. This partial closure of the inguinal canal also prevents the viscera from protruding into the scrotum. Such a protrusion of viscera into the scrotum is called a scrotal hernia. The inguinal canal cannot completely close because the spermatic cord must pass .through the opening. The spermatic cord is made up of nerves, blood vessels, connective tissue, the vasa deferentia, and the cremaster muscle, all surrounded by the tunica vaginalis. The cremaster muscle connects to the top of the testes and aids the tunica dartos in the thermoregulatory mechanism of the testes.

The Testes

The two testes (testicles) are the primary sex organs of the male. The testes are sometimes referred to as the gonads, although the term gonad refers to the primary sex organ of either sex. The testes function in producing the male gametes and in secreting the male sex hormone testosterone. Castration is the removal of the testes from the scrotum. This procedure removes the source of testosterone and the source of sperm. Castration is an effective method of sterilizing a male.

Structurally, the testis is made up of a thin-walled capsule of connective tissue packed with a network of small tubules and interstitial cells. The outer wall of the testis is called the tunica albugenia (see Fig. 9-3). The testicular capsule varies in length from 1 to 4 mm in cats and dogs to 100 to 140 mm in stallions and bulls. The diameter of the testis is usually about one-half to two-thirds of the length. A testis from a boar might measure 130 mm in length and 70 mm in diameter and weigh from 250 to 300 g. Inside the capsule of the testis are a number of compartments formed by partitions of connective tissue called septa. Extending through the central core of the testis is the mediastinum. The mediastinum is

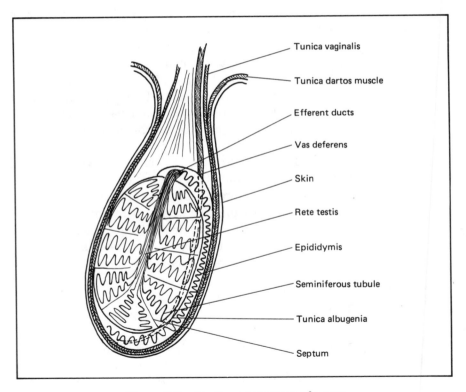

Figure 9-3   Testis and scrotum of a ram.

similar to a hallway that extends through the center of the testis leading to the several rooms or compartments formed by the septa.

The network of tubules begins inside the compartments of the testis with the seminiferous tubules. In a bull testis, the semiferous tubules are about 0.1 to 0.3 mm in diameter and from 50 to 100 cm in length. It has been estimated that the total length of seminiferous tubules in a single bull testis may be as much as 15,000 ft. One or more tubules are located inside each of the compartments of the testis. The tubules are very tightly packed into the testis and have a highly coiled and folded appearance. They are often referred to as convoluted seminiferous tubules. Figure 9-4 is a photograph of the testes of a bull, one of which has been sectioned.

Each compartment of a testis with the seminiferous tubules contained inside is called a lobe or lobule of the testis. In the spaces surrounding the tubules inside each lobe are found what are called interstitial cells or interstitial tissue. These cells are also called cells of Leydig and are the source of the male sex hormone. Also within the interstitial spaces are found connective tissue and blood vessels.

The seminiferous tubules lead into the mediastinum and continue as the rete testis. The rete testis is a network of straight tubules leading from the lobes of the testis down the mediastinum to the

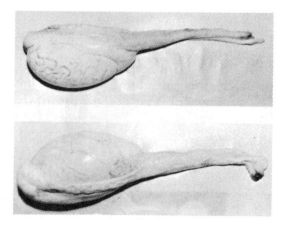

Figure 9-4   Testes of a bull removed from the tunica vaginalis. [H. Joe Bearden and John W. Fuquay, *Applied Animal Reproduction.* (Reston, Virginia: Reston Publishing Company, Inc., 1980) p. 30]

single opening at the end of the testis. The many tubules of the rete testis connect to about 13 to 15 short tubules called the efferent ducts. The efferent ducts all join together to form the single tube called the epididymis at the opening to the testicular capsule.

THE REPRODUCTIVE TRACT
AND ASSOCIATED STRUCTURES

The male reproductive tract of mammals consists of the urethra, vasa deferentia, and epididymides. Technically, it also includes the network of ducts located inside the testes, which were discussed above. In male birds the reproductive tract excludes the urethra, since birds do not possess such an organ.

The Epididymis

Each epididymis is highly folded or convoluted and situated inside a sac or pouch located next to each testis. Figure 9-3 illustrates the relationship between the epididymis and the testis on a bull. The epididymis is a long cordlike duct that gives the appearance of having been stuffed into the epididymal sac. It is relatively long when compared to the length of the vas deferens. In bulls and boars, the length of the epididymis has been found to approach 200 ft. This duct is generally divided into three regions known as the head, body, and tail. The head is the end nearest the connection to the testes. The tail of the epididymis connects to the vas deferens.

The epididymis functions in at least four ways. It is a storage place for sperm cells. As sperm are produced in the testis they are forced into the epididymis. If the sperm are not removed from the epididymis by ejaculation for long periods of time, some cells in the tail of the epididymis may be broken down and reabsorbed.

An obvious function of the epididymis is that of transportation of sperm cells from the testis to the vas deferens. The sperm must also spend some time in the epididymis before they become capable of fertilization. This is called the maturation function. Sperm cells removed from the head of the epididymis have been found to be incapable of fertilization. Those in the tail of the epididymis are viable. Much of the fluid that may have been secreted in the testis to aid in carrying the sperm cells into the epididymis is reabsorbed in the epididymis.

The Urethra and Vas Deferens

The urethra is as much a part of the reproductive system in male mammals as it is a part of the urinary system. It receives its name from its function as a transmission duct for urine. The urethra leads from the neck of the urinary bladder through the central core of the penis and terminates at the external or distal end of the penis. The external opening is called the urethral orifice. Birds do not possess a urethra or urinary bladder.

In mammals, two vasa deferentia connect to the urethra near the neck of the urinary bladder. Each vas deferens leads from the urethra downward through the abdominal cavity, through the inguinal canal, and into the scrotum where it joins an epididymis (refer to Fig. 9-1). In birds, the vasa deferentia lead from the epididymides and open into the cloaca. In some species of mammals the urethral ends of the vasa deferentia are somewhat enlarged in diameter. Each such enlarged region is called an ampulla. Ampullae are present in rams, bulls, and stallions. They are poorly developed in dogs and absent in boars.

In addition to castration, another method sometimes used to sterilize males is the vasectomy. This is a procedure whereby the vasa deferentia are cut to prevent the sperm from passing from the testes to the urethra. The animal continues to secrete testosterone and thus will continue to exhibit sex drive. Spermatogenesis continues in the testes, but since the sperm cells cannot be removed from the epididymis, they are reabsorbed. This sterilization without loss of sex drive is the same effect as obtained from congenital bilateral cryptorchism. Castration, however, results in loss of both sex drive and the ability to produce viable sperm.

The Accessory Glands

The accessory glands of the male reproductive system produce the fluid portion of the semen. This fluid, called seminal plasma, not only provides the means of transport for the sperm cells, it also provides nutrients, a buffering mechanism to control pH, and other means of protection and support. The accessory glands of mammals include the ampullae of the vasa deferentia, the prostate, the vesicular glands, and the Cowper's glands. The Cowper's glands are also called the bulbourethral glands. Vesicular bodies similar to ampullae are present in the vasa deferentia of male fowl. They do not possess the other accessory glands found in mammals. The Cowper's, prostate, and vesicular glands are associated with the urethra in mammals. The relative sizes of the glands vary among species but their relative locations are similar.

The vesicular glands are paired and open into the urethra in the same region as do the vasa deferentia at the neck of the urinary bladder. These glands are present in bulls, boars, rams, stallions, and rats, but are absent in dogs, cats, and rabbits. At one time the vesicular glands were called seminal vesicles because of the mistaken belief that they were storage vessels for semen. In some species of insects and other lower forms of animal life, the seminal vesicles do function as storage vessels for semen.

The prostate gland partially surrounds the urethra and releases its secretions through several openings. In some species, the prostate appears to be paired. Inflammation or infection of the prostate sometimes causes problems in older men and dogs. The Cowper's glands are much larger in boars than in bulls and rams and account for a large part of the added volume of seminal plasma produced by boars. Cowper's glands are present in all domestic mammals except dogs.

The Penis

The principal function of the penis, or copulatory organ, is to transfer semen from the male to the reproductive tract of the female. Two types of structures exist among domestic mammals with regard to the penis. One type is called the fibroelastic penis. It is rather rigid even in the nonerect state. This type of structure is common to bulls, rams, and boars, and has an S-shaped fold in it called the sigmoid flexure which straightens out during erection and copulation. Inside the penis and surrounding the urethra is a kind of spongy tissue called cavernosum. The cavernosum tissue is filled with many small spaces or cavities that become filled with blood during erection. The second kind of penis found in domestic animals is the vascular type common to dogs, stallions, and most other mammals. The vascular penis has no sigmoid flexure but is composed of cavernosum tissue, also. It is not nearly so rigid as the fibroelastic type when in the nonerect state.

When a male becomes sexually excited, the root of the penis is drawn tightly against the pelvis. This pressure is sufficient to cause the veins leaving the penis to collapse, but not the arteries that enter the penis. The blood continues to enter the penis by way of the arteries and begins to fill the many spaces within the cavernosum tissue. This engorgement of the penis with blood results in erection. The penis must be erect for semen to be successfully transferred to the female.

The pressure of the blood inside the fibroelastic penis causes the sigmoid flexure to straighten and the penis to extend from the prepuce or sheath. The vascular penis of horses and other mammals attains the needed length for copulation purely from the enlargement of the cavernosum tissue. Usually, after ejaculation, the blood pressure in the penis subsides and the penis returns to its nonerect state. A retractor muscle is present in the farm animal species to aid in retracting the penis into the sheath. This muscle is not present in smaller mammals.

SEMEN AND SPERM PRODUCTION

Semen and Seminal Plasma

Semen is composed of sperm cells suspended in a fluid called seminal plasma. The seminal plasma is secreted by the accessory glands and mixes with the sperm cells at the time of ejaculation. A small amount of the plasma is produced by the vasa deferentia to aid in the movement of the sperm cells into the urethra. The bulk of the seminal plasma in most domestic mammals is produced by the vesicular glands, the prostate, and the Cowper's glands and emptied into the urethra just prior to and during ejaculation.

The volume and concentration of semen produced by domestic animals is presented in Table 9-1. The variation observed within a given species is quite large. Domestic animals can be grouped into two categories based upon volume and concentration of semen. Stallions and boars produce the highest volume of semen, but their semen contains the lowest concentration of sperm. Dogs, bulls, and rams are similar in that they produce a relatively low volume of semen containing a relatively high concentration of sperm. The result is that the total number of sperm produced per animal per ejaculate varies from about 2 to 50 billion. Notice that, based upon these comparisons, the semen characteristics of men and cocks do not fit into the "high-low" category of stallions and boars nor into the "low-high" category of the other domestic animals. They are in a category of their own with low volume and low concentration.

Spermatogenesis

Sperm cells are produced in the seminiferous tubules from germ cells known as spermatogonia. The process is called spermatogenesis (see Fig. 9-5). Spermatogenesis begins at the time of sexual maturity,

TABLE 9-1. TYPICAL CHARACTERISTICS
             OF SEMEN FROM SELECTED ANIMALS

Animal	Volume per Ejaculate (mℓ)	Sperm Concentration (million per mℓ)	Total Sperm per Ejaculate (billions)
Boar	200 (150–250)	250 (100–400)	50 (15–100)
Stallion	100 (50–150)	150 (100–200)	15 (5–30)
Dog	7 (2–12)	1000 (500–2000)	7 (1–24)
Bull	6 (4–8)	1300 (1000–1600)	8 (4–13)
Ram	1 (0.5–1.5)	3000 (2000–4000)	3 (1–6)
Cock	1 (.05–2.0)	30 (1–60)	0.03
Man	3 (2–4)	100 (50–150)	0.3

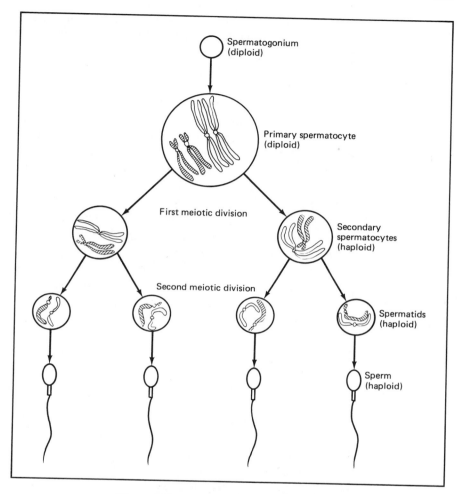

Figure 9-5   Diagram of spermatogenesis.

or puberty, and continues throughout most of the life of an animal until it becomes senile. Millions of spermatogonia exist inside the seminiferous tubules. They reproduce themselves by mitosis throughout the sexually active life of the male. This is to insure that the supply of these germ cells will never become exhausted.

The process of spermatogenesis can be divided arbitrarily into four phases to simplify the explanation. In the first phase the spermatogonium undergoes a number of changes in preparation for the first meiotic division. (The two types of cell division, mitosis and meiosis, were explained in some detail in Chapter 5.) The amount of cytoplasm in the cell increases, causing the cell to become much larger. The chromosomes duplicate themselves, resulting in a doubling of the genetic material within the cell. The pairs of homologous chromosomes seek out one another and lie side by side until the first division occurs. The membrane surrounding the nucleus disappears. During the time most of these changes are taking place, the cell is referred to as a primary spermatocyte. The first phase of spermatogenesis, then, is the formation of the primary spermatocyte from a spermatogonium.

The second and third phases involve a series of two cell divisions. The first division results in the formation of two cells called secondary spermatocytes. This division is sometimes called reduction division because the chromosome number of the original cell is reduced by one-half. One chromosome from each pair of homologs is transferred to each of the two secondary spermatocytes produced by the first division.

Shortly after the first division has been completed, the second division occurs. The second division results in the formation of four spermatids from the two secondary spermatocytes.

The fourth phase of spermatogenesis results in the formation of the sperm cells from the spermatids. This part of spermatogenesis is sometimes referred to as spermiogenesis. Unfortunately this term can easily be confused with spermatogenesis. The transformation of the spermatid to a sperm cell involves the development of a tail and the loss of most of the cytoplasm of the cell.

The whole process of spermatogenesis occurs inside the seminiferous tubules. Figure 9-6 shows a diagram of a cross section of a seminiferous tubule. The cells nearest the outer wall of the tubule are spermatogonia and primary spermatocytes. As the cells undergo the changes that are a part of spermatogenesis they are forced toward the core or lumen of the tubule. The various spermatogenic cells are supported by other cells in the tubule called Sertoli cells, or nurse cells. Sperm cells probably receive nutrients from the Sertoli cells. When the sperm cells are finally formed, they are released into the lumen of the seminiferous tubule and move toward the rete testes.

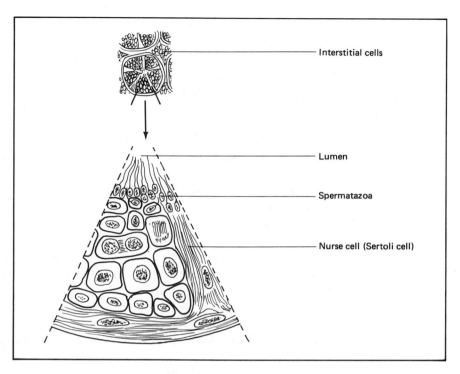

Figure 9-6 Cross section of mammalian seminiferous tubule showing spermatogenesis in various stages of development.

Sperm Structure and Physiology

The shape and structure of the sperm cells of most domestic animals
are very similar. The head varies in length from about 6 to 10 μ, the
width from 5 to 8 μ, and the thickness from 1 to 2 μ. The midpiece
is about twice the length of the head and a little larger in diameter
than the tail. The tail is about 1 μ in diameter and varies in length
from 40 to 50 μ. The head of poultry sperm is about 15 to 20 μ
in length and much more slender in appearance than mammalian
sperm cells.

The head of the sperm cell is covered with a caplike membrane
called the acrosome. About one-half of the volume of the sperm
head is made up of the nucleus. The nucleus contains one-half the
usual number of chromosomes found in diploid body cells. For

this reason sperm are referred to as haploid. A diagram of a mammalian sperm is shown in Figure 9-7.

Occasionally, sperm cells are referred to by the term spermatozoa. The singular form of this word is spermatozoon. However, this terminology is not used as frequently as it once was. In much the same way that the word testicle has given way to testis, spermatozoon has given way to sperm. A sperm cell is a highly specialized, nondividing cell whose only function is to carry the male half of the inheritance to the female. Its ability to fertilize the female ovum is highly dependent upon its motility. The tail of the sperm cell is largely responsible for its motility.

The ability of the sperm to fertilize is stimulated by secretions of the epididymis. Sperm removed from the head of the epididymis have not attained fertilizing ability. By the time they reach the tail of the epididymis, however, they have attained that ability. Sperm cells can remain alive in the tail of the epididymis for up to 60 days. After prolonged periods of sexual inactivity, sperm cells in the epididymis will degenerate. For this reason, it is wise to collect

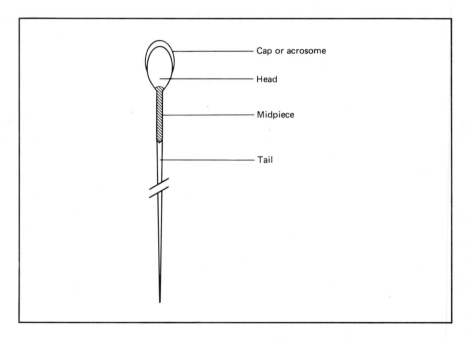

Figure 9-7   A typical mammalian sperm cell.

several samples of semen from a male before he is used for service if he has not been sexually active for several months.

Once the sperm have entered the female tract in mammals, their life becomes shortened. In the reproductive tract of the cow and ewe, the sperm will remain fertile for only 1 or 2 days. In mares, sperm may survive for as long as 6 days. In the reproductive tract of female poultry, the sperm may remain fertile for up to 30 or 40 days. By contrast, the queen honeybee mates only once in her life and may store viable sperm in her body for several years.

HORMONES OF REPRODUCTION IN THE MALE

Reproduction in the male is controlled directly by at least three hormones, two from the pituitary gland and one from the testes. The two hormones of the pituitary are called gonadotropins because they stimulate the gonads. They are named for their functions in the female. Follicle-stimulating hormone (FSH) is a product of the anterior lobe of the pituitary gland. It stimulates the seminiferous tubules to produce sperm. In other words, it stimulates spermatogenesis.

The second gonadotropin is also produced by the anterior pituitary gland. It is commonly called luteinizing hormone (LH). It stimulates the interstitial cells (cells of Leydig) of the testes to produce the male sex hormone testosterone. In the male it is sometimes referred to as interstitial-cell-stimulating hormone (ICSH).

Testosterone has a number of functions in the male. It stimulates the growth, development, and activity of the male reproductive organs and accessory glands. It stimulates the production of sperm, so in this function it works with FSH. Male sex drive, or libido, is stimulated by testosterone. Testosterone also is partially responsible for development of the secondary sex characteristics of the male such as the crest of the bull, the prominent comb development and coloring of male birds, the deeper sound of the voice, larger body size, and heavier muscling.

The relative degree of masculinity in males is due to the relative levels of male and female sex hormones in the body. In a normal male the levels of testosterone will be relatively high; however, some of the female sex hormone, estrogen, will also be present.

Masculine traits, or secondary male sex characteristics, are due to the relative levels of testosterone and the testosterone-estrogen ratio. Some males that appear to be very masculine may have a very high level of testosterone or an exceptionally low level of estrogen, or both. The narrower the ratio between these two hormones becomes, the less masculine the male will appear. Sometimes in cases where excessive amounts of estrogen are secreted by the adrenal glands or the testes, males may show many of the secondary female characteristics. These characteristics are also caused by testes that fail to secrete enough testosterone to allow the male characteristics to develop properly.

STUDY QUESTIONS AND EXERCISES

1. Define each of the following terms:

acrosome	prostate
ampulla	puberty
castration	rete testis
cavernosum	semen
cloaca	seminiferous tubules
copulation	septum
cryptorchid	Sertoli cells
efferent ducts	sigmoid flexure
ejaculation	spermatid
epididymis	spermatogenesis
erection	spermatogonium
gonad	spermatozoon
gonadotropin	testis
hernia	testosterone
inguinal canal	tunica albugenia
interstitial cells	tunica dartos
Leydig cells	tunica vaginalis
libido	urethra
mediastinum	vas deferens
monorchid	vasectomy
penis	zygote

2. What is the general function of the male reproductive system?

3. Trace the path of a sperm cell from the site of its formation to the outside of the animal.

4. Compare the structures of the male reproductive systems of a typical mammal and a cock (rooster).

5. Compare the structure of the penis of the stallion to that of the bull.

6. List the functions of the epididymis.

7. Name the accessory sex glands and indicate their function.

8. Describe the process of erection of the penis of a bull.

9. Name and locate the two muscles that help control the temperature of the testes.

10. Explain the origin of the tunica vaginalis.

11. Why are the testes of mammals suspended outside the body rather than inside the body cavity as in birds?

12. Compare castration and vasectomy as methods of sterilization.

13. Compare the several species of domestic animals as to volume and concentration of semen.

14. How does the volume and concentration of human semen compare to the various domestic animals?

15. Explain the mechanism and significance of spermatogenesis.

16. How does a sperm cell compare to other cells of the body?

17. How long can sperm cells live in the reproductive tract of females compared to the length of life in the epididymis.

18. Compare the length of fertilizable life of sperm cells in the female reproductive organs of the various animal species.

19. Compare the functions of FSH and LH (ICSH) in the male animal.

20. List the functions of testosterone.

10

Artificial Insemination

One of the most important techniques that has been devised to further the genetic improvement of animals is artificial insemination. Almost one-half of all dairy cattle in the United States are bred artificially. The use of artificial insemination (A.I.) has shown varying degrees of success in sheep, swine, beef cattle, horses, poultry, and even humans. In the dairy cattle industry the use of A.I. has made it possible to use fewer, more superior sires on many more cows than would have been possible with natural breeding methods. Commercial turkey breeding relies almost totally upon artificial insemination which results in a significant increase in conception rates compared to natural service.

SEMEN COLLECTION

Semen may be collected from bulls, rams, boars, and stallions by the use of an artificial vagina (A.V.). The general construction of an artificial vagina for collecting semen from a bull is shown in Figure 10-1. In general it is composed of a rubber tube surrounded by a water jacket filled with warm water. At one end is attached a rubber funnel which is in turn attached to a collecting tube. The collection can be made as the male attempts to mount the female. The penis is guided into the artificial vagina. This method of semen collection duplicates the natural process of copulation for the male more closely than any other method. For this reason it is the preferred method of collection. Males can be trained to mount dummies, steers, or cows not in heat. Bulls and rams produce a relatively small volume of semen and require only a minute or two to mount, ejaculate, and dismount. Because stallions and boars produce a larger volume of semen, it takes longer to collect from them. This is especially true of the boar from which it sometimes requires five or ten minutes to complete the collection process.

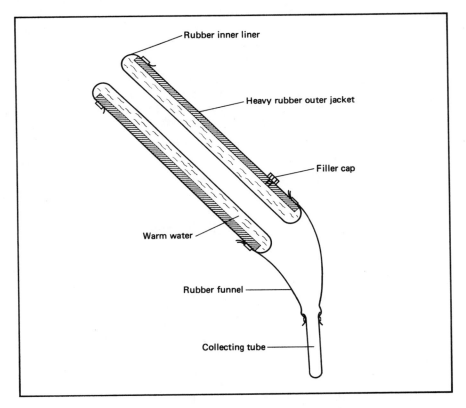

Figure 10-1   Diagram of an artificial vagina.

Another method of collection may be necessary for males that refuse to serve the A.V. or when injuries make it impossible. The most common alternate method involves the use of the electroejaculator. An electrical probe is placed in the rectum. The animal is subjected to brief periods of low voltage current which stimulates the nerves in the region of the base of the penis that in turn control erection and ejaculation. Bulls and rams respond quite well to electrical stimulation. This method is less satisfactory for use with boars. One type of electroejaculator is shown in Figure 10-2.

If it is not convenient or possible to use an A.V. or an electroejaculator, it is possible to collect from bulls by mechanical stimulation. Pressure can be applied through the rectum to the vesicular glands, the vasa deferentia, and ampullae. Since this will not usually cause the bull to have an erection, however, only small amounts of semen may be obtained by this method. Moreover, the semen may become contaminated with bacteria of other materials from the prepuce. The massage or "milking" method is the usual procedure for collecting semen from turkeys and chickens.

It is possible to collect semen from rams and bulls as often as once a day for several weeks without any serious depletion of semen volume and sperm numbers. Generally, collections are made from bulls by artificial breeding associations about three or four times a week. Because boars and stallions produce a larger volume of semen per ejaculate, it is best to collect from them a little less often, maybe two to three times a week.

Figure 10-2 Electroejaculator showing the probes used for a bull and a ram. [H. Joe Bearden and John W. Fuquay, *Applied Animal Reproduction*. (Reston, Virginia: Reston Publishing Company, Inc., 1980) p. 154]

Once the semen is collected it should be protected from rapid changes in temperature. An increase in temperature results in an increase in the metabolic activity of the sperm thus decreasing its life span. If the temperature goes above 50°C, there is an irreversible loss in motility and viability. The metabolic rate of the sperm approximately doubles for every 10°C increase in temperature. On the other hand, decreasing the temperature of the sperm decreases their metabolic activity and extends their life-span. It is important to avoid shock, however. When sperm are quickly cooled to about the freezing point they suffer an irreversible loss of viability. Cold shock can be avoided by cooling the semen slowly, particularly in the region between 15° to 0°C. When sperm are cooled in this manner and then rewarmed, their motility and viability are restored. The exception is semen from boars. Boar semen should not be cooled below 15°C. The semen of rams and stallions can be stored for about a day at 5°C. Bull semen can be kept up to four days at this temperature. Boar semen can be stored for about two days, but at a temperature no lower than 15°C.

In order to evaluate the quality and viability of semen, a few tests have been developed. Properly collected semen of high quality should have a milky, opaque appearance, and should be quite viscous. This indicates a high sperm concentration. The volume should be representative of the species. The relative motility can be estimated by a trained operator by placing a drop of semen on a slide and observing it through a microscope. A good sample will show about 70 percent motility.

The proportion of live to dead sperm cells can be determined by staining a sample of semen on a microscope slide with a special dye. The dye will be absorbed by the dead sperm but not by the live ones. Microscope slides of sperm can also be observed for variations in morphology. High-quality semen will contain few abnormal sperm. Large numbers of abnormal sperm, up to 30 percent, would indicate that the semen is of poor quality. Abnormalities of sperm include such features as bent, missing, coiled, and poorly formed tails, abnormal midpieces, and small, enlarged, and odd-shaped heads. These abnormalities could be brought about by nutritional or hormone problems, by cold or heat shock, or by the effects of exposure to sunlight.

The number of sperm per unit volume, or concentration, can be determined in any one of several ways. One common method is to determine the turbidity of the sample using a photoelectric colorimeter. This device measures the amount of light that passes through the sample. The greater the number of sperm in the sample, the greater the turbidity and thus the less amount of light that will pass through the sample. Two methods that are commonly used to determine the number of red blood cells in a blood sample can also be used to determine sperm concentration. One involves the use of a hemocytometer. The hemocytometer is a specially designed microscope slide with calibrated markings. The number of sperm cells per given volume can be counted by observing them through a microscope. The second of these methods involves the hematrocrit technique. A small amount of semen is placed inside a small glass tube and placed in a specially designed centrifuge. The length of the column of packed cells is expressed as a percentage of the total volume of semen in the tube. From this, an estimate can be made of the number of sperm cells present in the sample.

SEMEN EXTENSION AND STORAGE

In natural mating one ejaculate is used to inseminate one female. Various sources indicate that from 5 to 12 million live sperm in 1 mℓ of diluted semen is all that is needed for satisfactory conception rates in females inseminated artificially. From the information presented in Table 9-1, it can be seen that considerably more sperm than that is found in one ejaculate from domestic animal males. The main objective in extending semen is to increase the volume of semen so that many more females can be bred to one male. The actual number of females that can be inseminated by one ejaculate is dependent upon the volume of semen, the sperm concentration, and the percentage of motile sperm. The following example shows how a typical sample of bull semen can be extended to inseminate a large number of cows. Assume that one ejaculate has a volume of 7 mℓ, a concentration of 1.2 billion sperm per mℓ, and a motility rating of 70 percent. One mℓ of semen would contain 0.70 × 1,200,000,000 or

840,000,000 live sperm. If we assume that 8,000,000 sperm are needed in a 1-mℓ sample of diluted semen then 1 mℓ of the fresh semen contains enough sperm to inseminate 840,000,000 ÷ 8,000,000 = 105 cows. The 7mℓ ejaculate contains enough live sperm to inseminate 7 × 105 = 735 cows. If four collections were made from this bull each week, he could be bred to almost 3000 cows per week or 150,000 cows in one year. Depending upon the volume, concentration, and sperm motility, one ejaculate from a bull may be used to inseminate from 100 to 1000 cows. Due to different levels of sperm production and the numbers required for a successful service, the numbers of females that can be bred artificially to other domestic animal males is somewhat less than that for the bull. About 100 ewes, 20 sows, 10 mares, and 10 hens can be inseminated with one ejaculate from the males of their respective species.

In order to provide the ejaculate with sufficient volume to permit multiple inseminations, semen extenders are used. In addition to extending the volume of semen, a good semen extender must also serve other functions. It should provide nutrition, protect against changes in pH, prevent bacterial growth, protect against cold shock, and otherwise maintain the viability of the sperm over long periods of time. Most extenders contain either egg yolk or milk or a combination of both. In addition most extenders also contain glycerol, buffers, and antibiotics to provide the above characteristics.

After the semen has been mixed with the extender, it may be stored at 5°C for one to two days depending on the species. Artificial insemination of sheep, swine, horses, and turkeys makes use of fresh semen or semen that has been stored for only a few days. Very few cattle are now being inseminated artificially with fresh semen. Nearly all A.I. studs are now freezing bull semen. Frozen semen can be used successfully in cattle, horses, goats, dogs, and humans.

The various methods of freezing semen include mechanical freezers, dry ice and alcohol, liquid carbon dioxide, and liquid nitrogen. Liquid nitrogen freezers are the most common method now used (Fig. 10-3). Semen can be stored for several years at a temperature of −196°C. The temperature of the semen is usually lowered at about 3 degrees per minute until it reaches −15°C. It can then be lowered more rapidly down to −80°C and then transferred to the liquid nitrogen freezer at −196°C.

Semen is normally stored in ampules containing 0.5 to 1.0 mℓ

Figure 10-3 Liquid nitrogen tanks used to freeze and store semen. [H. Joe Bearden and John W. Fuquay, *Applied Animal Reproduction*. (Reston, Virginia: Reston Publishing Company, Inc., 1980) p. 181]

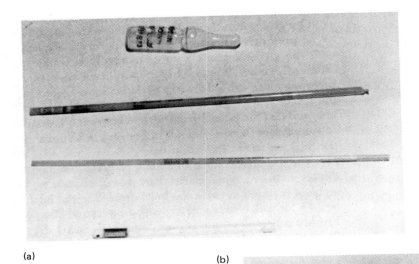

(a)

(b)

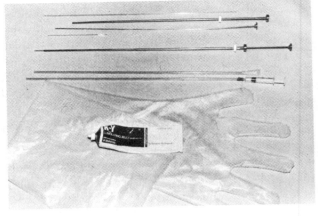

Figure 10-4 Materials used (a) in freezing and storing semen and (b) in artificial insemination. [H. Joe Bearden and John W. Fuquay, *Applied Animal Reproduction*. (Reston, Virginia: Reston Publishing Company, Inc., 1980) pp. 178 and 191]

246

of diluted semen or in straws containing 0.25 to 0.5 mℓ of semen. Frozen semen should not be thawed until just prior to insemination. It can be thawed by placing it in ice water for about ten minutes. Semen ampules, straws, and other inseminating equipment for cattle are shown in Figure 10-4.

INSEMINATING THE FEMALE

One of the first problems associated with artificial insemination in farm mammals is the detection of estrus. Cows in estrus (heat) will attempt to mount other cows and allow themselves to be mounted. They are restless and will bawl more than usual. Sows in estrus will become immobile when weight is applied to their backs. Acceptance of the male is probably the best indicator that a female is in estrus, but this procedure does not work very well with artificial insemination. This principle can be utilized, however, by the use of vasectomized males as "heat checkers." The problem of estrus detection is of particular importance in beef cattle because they have an unusually short estrus and are not usually observed as closely as are dairy cattle and other species.

Ovulation in cattle, sheep, and swine normally occurs during the second day after the beginning of estrus. Insemination should take place very near the time of ovulation. In cattle this means that insemination should occur several hours after the cow ceases to show signs of estrus.

In order to aid in the process of estrus detection and increase the efficiency of artificial insemination, a procedure called estrus synchronization can be practiced. The procedure involves the adjustment of the estrous cycles of a large group of females so they will all show estrus on the same day. They can then all be bred by artificial insemination at the same time. For example, cows can be fed or injected with progesterone for a period of about two weeks. The progesterone will result in the inhibition of the development of ovarian follicles. Following the withdrawal of progesterone, the cows will normally come into heat within a few days. Another procedure involves the use of a substance called prostaglandin. It acts by causing the corpora lutea to regress, followed by estrus a few days later.

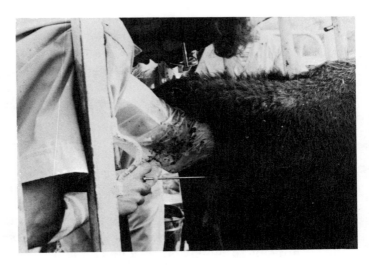

Figure 10-5 Inseminating a cow.

Estrus synchronization is especially useful in beef cattle because it reduces the time involved in observing cattle for signs of estrus. Another advantage of estrus synchronization in farm animals is that it results in a much shorter calving, lambing or farrowing period.

The insemination of a cow is shown in Figure 10-5. The inseminator puts one arm inside the rectum of the cow to manipulate the cervix as the catheter or inseminating pipette is guided through. Because the volume of semen is lower and contains fewer cells than semen deposited naturally by a bull, it is necessary that the semen be deposited inside the uterus.

STUDY QUESTIONS AND EXERCISES

1. Define each of the following terms:
   A.I.   ejaculate
   artificial vagina   electroejaculator
   A.V.   hemocytometer

2. How extensive is the use of artificial insemination techniques in domestic animals?

3. Compare the artificial vagina and the electroejaculator as means of semen collection from the various species of domestic animals.

4. How often can semen be collected from various domestic animals as routine practice?

5. What is the relationship between both high and low temperatures and length of sperm life?

6. What is cold shock and how can it be prevented?

7. What are some characteristics of a good sample of freshly collected semen?

8. How many live sperm should be inseminated artificially to insure that fertilization will occur?

9. How many units of semen can be obtained from a fresh 8 mℓ sample of bull semen that contains 950,000,000 sperm per mℓ and only 65 percent motility? Assume that a unit of semen must contain 7 million live sperm cells. (Answer: 705 units.)

10. If collections were made from the bull in the above problem three times each week, how many cows could he theoretically breed in one week? In one year? (Answer: 2115 cows per week or almost 110,000 cows per year.)

11. What are some of the functions of a good semen extender?

11

Nutrients

Animal nutrition deals with the supplying of nutrients to animals and the digestion, absorption, and utilization of these nutrients. The nutritionist must know something about the nature and quantity of nutrients required to meet the needs of the animal for maintenance, growth, reproduction, lactation, and work. Just as it requires a knowledge of genetics to be a good animal breeder, it requires a knowledge of nutrition to be a good animal feeder. In order to obtain the maximum and most efficient production from domestic animals, the proper nutrients in adequate amounts must be supplied in the right kind of feeding program. The knowledge and proper application of the principles of breeding, disease control, economics, and general management are of little or no value if animals do not obtain the proper quality and quantity of nutrients.

Some nutrients, such as starches and proteins, are rather complex. Others, such as the mineral elements, are rather simple. The more complex nutrients usually must be chemically broken down, or digested, into simpler compounds in order to be properly absorbed and utilized. Examples of this kind of nutrient include the proteins, fats, and some of the carbohydrates. Other nutrients, such as water and most of the minerals and vitamins, may be absorbed from the digestive tract without digestion. The digestive system serves as a place to receive food materials, carry out the necessary digestion, and make the simplified nutrients available for absorption into the blood of the animal.

CARBOHYDRATES

The carbohydrates are a large group of organic compounds that include sugars, starches, and cellulose, among others. Carbohydrates are composed of the elements carbon, hydrogen, and oxygen and are the most abundant group of organic compounds on earth. They are basically plant substances although small amounts of certain carbohydrates are found in animals. Plants synthesize carbohydrates through the process of photosynthesis. Beginning with carbon dioxide from the atmosphere, and water, the chlorophyll of the plant manufactures sugars using energy from the sun. The sugars are then used to build large molecules of starch and cellulose and other large carbohydrates in the plant. Following is a simplified diagram of photosynthesis:

$$6CO_2 \quad + 6H_2O \longrightarrow C_6H_{12}O_6 + \quad 6O_2$$

carbon dioxide + water + sun energy = sugar + oxygen

Note that, whereas animals breath in oxygen and expire carbon dioxide, plants carry out the reverse process.

In plants, carbohydrates serve basically a twofold function. Some are used to provide a source of energy to growing and developing plants during times when they are not synthesizing carbohydrates. Others are used to form the skeleton or structural components of the plant. Only a few carbohydrates are found in animals—glucose in the blood, glycogen in the liver and muscle, and lactose in milk, to name a few. Animals make use of carbohydrates, which they obtain primarily from plants, as a source of energy. In fact, carbohydrates make up the single largest source of food for domestic animals.

The Sugars

Sugars are usually divided into two categories, the monosaccharides and the disaccharides. The monosaccharides, or simple sugars, are made up of single sugar units. They are used as building blocks for larger carbohydrates. Only a few of the monosaccharides are found in nature in the free form. Glucose is the most common of the sugar building blocks of nature. It is found in the free form in the blood and in certain ripe fruits. Starches and cellulose are made up of hundreds of glucose units. Another of the monosaccharides is fructose, sometimes called fruit sugar because it is also found in many ripe fruits. A third simple sugar is galactose which is almost never found in the free form. It is a component of the disaccharide sugar called lactose.

Disaccharides are made up of two monosaccharides connected by a chemical bond. These sugars are sometimes called compound sugars. Three such sugars are common in nature; they are maltose, sucrose, and lactose. Maltose is made up of two units of glucose. It should be stressed that maltose is not just a mixture of glucose— it is chemically a completely different sugar. Maltose is commonly found in the sprouting tips of growing plants. Malt, which is made from barley sprouts, is a common source. Lactose is the sugar found in milk. It is known to exist almost nowhere else in nature. Lactose is made of a unit of glucose and one unit of galactose.

Sucrose is the most common of the naturally occurring sugars that exist in the free form. It is found in the saps of most plants. The most common commercial source is sugar cane. Sugar beets and sugar maple are two other common sources. The following illustration shows the structures of glucose and fructose and how they combine to form sucrose:

glucose   fructose

$$CH_2OH \qquad CH_2OH$$

sucrose

Two characteristics of sugars are that they are soluble in water and are sweet. Some sugars are sweeter than others. Comparing the sweetness of various sugars to that of sucrose, taste test panels have determined the following relative degrees of sweetness for the six common sugars:

Fructose	174
Sucrose	100
Glucose	72
Maltose	32
Galactose	32
Lactose	16

Notice that fructose is about three-fourths sweeter than sucrose, whereas milk sugar is only about one-sixth as sweet.

Starch and Cellulose

Carbohydrates made up of hundreds of monosaccharide building units are called polysaccharides. The simple sugars are connected in chainlike fashion in much the same way as the two monosaccharides are joined to build sucrose. The two most common polysaccharides are starch and cellulose. Starch is manufactured by plants and stored in the seeds, roots, and other parts of some plants to be used by the plant as a source of energy. Sprouting seeds call upon their starch stores to help the plant develop until it can manufacture enough chlorophyll to begin photosynthesis.

Starch is rather easily digested by animals and provides the major source of energy to monogastric animals. On the other hand, cellulose is more complex than starch and is much more difficult to digest. In fact, it can only be digested by microorganisms in the rumen and the cecum.

FATS

Fats are generally smaller in molecular size than starches and protein. They are not soluble in water, but will dissolve in organic solvents such as ether, acetone, chloroform, and xylene. Because of their solubility characteristics, they are included in a large group of compounds known as lipids. Lipids include another large group of compounds called steroids. Cholesterol, bile, and sex hormones are examples of steroids. Chemically, steroids have very little in common with fats.

Fats are composed of subunits called fatty acids and glycerol. A typical fat molecule might look something like this:

$$
\begin{array}{c}
\qquad\qquad \overset{\displaystyle H}{|}\ \ \overset{\displaystyle O}{\|} \\
\qquad\qquad HC-O-C-R'' \\
\overset{O}{\|}\qquad\qquad | \\
R'-C-O-CH\qquad \overset{O}{\|} \\
\qquad\qquad | \\
\qquad\qquad HC-O-C-R \\
\qquad\qquad \underset{\displaystyle H}{|}
\end{array}
$$

The R-groups represent the rest of the fatty acid molecule. Each R-group will contain about 3 to 19 additional carbon atoms. The three-carbon center of the molecule is the part contributed by glycerol. Because a fat molecule obtains from 90 to 95 percent of its composition from the three fatty acid molecules, it tends to take on some of the characteristics of those fatty acids. Such things as molecular weight, melting point, and degree of saturation of the individual fatty acids affect the characteristics of a particular fat molecule.

Nutritionally, fats are used by animals as an energy source. The amount of energy that a fat or carbohydrate can contribute is related primarily to the amount of carbon and hydrogen that it contains. Because fats contain so much more carbon and hydrogen than do the carbohydrates, they can contribute much more energy. A typical fat molecule might have the formula $C_{55}H_{102}O_6$. This molecule of fat would contain about 89 percent carbon and hydrogen and only 11 percent oxygen. A starch fragment with about the same number of carbons would have the formula $C_{54}H_{92}O_{46}$ which contains about 50 percent oxygen and 50 percent carbon and hydrogen. The amount of energy in the starch fragment would be considerably less than that in the fat molecule. Studies have shown that fats contain about two and one-fourth times the amount of total energy found in carbohydrates.

Fatty Acids

The molecular size of fatty acid molecules varies from molecules with 2 carbons, as in acetic acid, to those with 20 or more carbons. Generally the fewer the number of carbons in a fatty acid, the lower its melting point. Acetic acid (C_2), butyric acid (C_4), and caproic acid (C_6) are liquids at ordinary room temperatures. Because of their relatively low melting points, they volatilize easily and are sometimes called volatile fatty acids (VFA). Acetic acid and butyric acid along with the three-carbon propionic acid are produced by microorganisms in the rumen. Because of the unique way in which they are synthesized in plants and animals, the naturally occurring fats are almost always composed of an even number of carbons.

The larger saturated fatty acids are solids at ordinary temperatures. A saturated fatty acid is one that has all of the hydrogen atoms that the carbons can hold. The following illustration shows a fragment of a saturated fatty acid:

$$
\begin{bmatrix}
& H & H & H & H & H & H & H & H & \\
& | & | & | & | & | & | & | & | & \\
-\!\!\!& C & C & C & C & C & C & C & C &\!\!\!- \\
& | & | & | & | & | & | & | & | & \\
& H & H & H & H & H & H & H & H &
\end{bmatrix}
$$

If some of the hydrogen atoms were missing, some of the carbons would be connected by double bonds and would be called unsaturated, as in this illustration:

$$
\begin{bmatrix}
& H & H & H & & & H & H & H & \\
& | & | & | & & & | & | & | & \\
-\!\!\!& C & C & C & C & = & C & C & C & C &\!\!\!- \\
& | & | & | & & & | & | & | & \\
& H & H & H & H & & H & H & H & H &
\end{bmatrix}
$$

The most common saturated fatty acids found in natural fats are these:

Lauric acid $C_{12}H_{24}O_2$

Myristic acid $C_{14}H_{28}O_2$

Palmitic acid $C_{16}H_{32}O_2$

Stearic acid $C_{18}H_{36}O_2$

Unsaturated fatty acids have lower melting points and are usually liquids at room temperatures. The most common unsaturated fatty acids found in natural fats are the following:

Oleic acid $C_{18}H_{34}O_2$ (one double bond)

Linoleic acid $C_{18}H_{32}O_2$ (two double bonds)

Linolenic acid $C_{18}H_{30}O_2$ (three double bonds)

Most of the fatty acids used by the animal body to synthesize fat can be made from other fatty acids in the body. However, animals have difficulty synthesizing linoleic acid, linolenic acid, and the 20-carbon arachidonic acid, the latter having four double bonds. The requirements for these acids are so low, however, that the amounts of fat normally consumed by domestic animals will usually contain a sufficient supply of them. Because it is so important that these three fatty acids be present in the diets of animals, they are often referred to as essential fatty acids.

Natural Fats

Fats are found in both plants and animal tissues. All fats are composed of the glycerol core which is attached to three fatty acids. Theoretically, any combination of three fatty acids may be found in a particular fat molecule. Some molecules may have two or three of the same fatty acids in their makeup. Generally, fats from animal sources have a greater percentage of the saturated fatty acids than do fats from plant sources. As an example, consider the four fats shown in Table 11-1, two from plants and two from animals, and their

TABLE 11-1. FATTY ACID COMPOSITION
OF ANIMAL VERSUS PLANT FATS

Fatty Acid	Butterfat	Lard	Soybean Fat	Corn Fat
Palmitic	27%	32%	9%	7%
Stearic	13	8	4	2
Oleic	35	48	17	46
Lineolic	3	11	54	45
Total fatty acids	100%	100%	100%	100%

relative percentages of four fatty acids. Palmitic and stearic acids are saturated; oleic and linoleic acids are unsaturated. The percentages of saturated fatty acids in the fats of butter, lard, soybeans, and corn are approximately 60, 40, 20, and 10, respectively. Because fats tend to exhibit the characteristics of the fatty acids they contain, those from plant sources tend to be liquids at room temperature while those from animals tend to be solids. The plant fats are sometimes referred to as oils because they are usually liquids.

An interesting characteristic of fats is that in both plant and animal tissues, the molecules tend to cling together to form globules. Where fat molecules may be relatively small, fat globules can be very large. In some ways a fat globule could be thought of as a large popcorn ball with the individual kernels representing the fat molecules.

PROTEINS

Proteins are generally more complex in composition than fats and carbohydrates. Carbohydrates, no matter how large, will usually be made up of just one or two kinds of building units. Proteins are composed of smaller building units called amino acids. However, there are about 20 different amino acids that can be found in naturally occurring proteins. Proteins are composed of carbon, hydrogen, and oxygen as are the fats and carbohydrates. However, they also contain about 16 percent nitrogen. Nitrogen is a necessary component of protein and the amount varies only a little from protein to protein. Some sulfur is also found in proteins.

Because proteins are involved to a large extent in the structure of individual cells, much of the protein intake of animals goes to build new cells—particularly in the case of growing animals—or to replace worn-out cells. Many of the vital substances in the body fluids and cells of animals are made of protein, including some of the hormones, antibodies, and enzymes. If the protein consumed by an animal is in excess of what is needed to build and maintain the cells and vital substances, it can also be used to supply energy to the body. On a practical basis, however, this is usually not an economical way to supply energy since protein feeds are generally more expensive than fats and carbohydrates.

Proteins are made up of amino acids held together chemically in chainlike fashion. The particular number and sequence of the 20-odd amino acid building blocks varies with the tissues and cells that are responsible for the construction of the proteins. Proteins could be thought of as very large words with the amino acids being the letters that make them up. Just as two different words require a different number and sequence of certain letters, proteins require a certain number and sequence of certain amino acids. Not only are the proteins in different tissues within the same body different, proteins within the same tissues in different animals will be different. The more different the animals, the more their proteins will differ. That difference becomes even greater when we begin to compare animal proteins to plant proteins. The same 20 or so amino acids may be found, but the relative proportions of each will vary greatly.

Amino Acids

Amino acids contain carbon, hydrogen, oxygen, and nitrogen. The basic structure of all amino acids, with but a few exceptions, is

$$R-\underset{\underset{NH_2}{|}}{C}-C\underset{\diagdown OH}{\overset{\diagup\!\!\!\!\diagup O}{}}$$

The R-group varies from one amino acid to the next. Some contain sulfur, while some contain additional nitrogen or oxygen. All contain additional hydrogen, and all but one contain additional carbon. Animals cannot construct this basic unit from the individual elements of carbon, hydrogen, oxygen, and nitrogen. This means that they must meet their amino acid needs from the protein that is in their food. The protein is digested to yield the individual amino acids which are then absorbed into the blood. The blood carries the amino acids to the tissues where they are used to construct some particular protein. So, when we discuss protein needs of animals we are to a large extent discussing amino acid needs.

In some amino acids the part of the molecule represented by the R, is simple enough that many animals are capable of rearranging the structure to build another amino acid that may be in short supply in the food. Other amino acids are constructed in such a way that rearrangement of their structure cannot be handled by the animal at a rate fast enough to meet a particular need. It is important that the amino acids in this latter group be found in the food of the animal. If they are not, the animal cannot build the particular proteins that require those building blocks. Since it is essential that this group of amino acids be present in the protein that the animal eats, they are called essential amino acids. The term indispensable is sometimes used in referring to the essential amino acids. Those that can be constructed from other amino acids are then called nonessential or dispensable. These terms refer to the presence of the various amino acids in the diets. At the cell level, all of them are needed. In the diets of pigs and rats and most monogastric animals, the 20 common amino acids can be separated according to their dispensability as shown in Table 11-2. The degree of dispensability of the various amino acids is not clear cut in all animals. In humans, for example, histidine and arginine are nonessential. In chicks, glycine is essential. Methionine, an essential amino acid, contains sulfur. Because cystine and cysteine also contain sulfur, some of the methionine requirement can be met with these two amino acids. Because of their similarity in structure, part of the phenylalanine requirement can be met with tyrosine.

The subject of amino acid dispensability is not an important consideration in ruminant animals. The microorganisms in the rumen have the capability to synthesize from the basic elements all of the amino acids needed to build their own cellular structures. All that is necessary is that ample amounts of carbon compounds be made available along with a sufficient amount of nitrogen. It is usually best that these requirements be met with ample amounts of protein. The particular amino acid content of that protein is not important.

TABLE 11-2.   ESSENTIAL AND NONESSENTIAL
              AMINO ACIDS
              IN MONOGASTRIC ANIMALS

Essential (Indispensable) Amino Acids	Nonessential (Dispensable) Amino Acids
Phenylalanine	Glycine
Valine	Alanine
Tryptophan	Serine
Threonine	Cysteine
Isoleucine	Cystine
Methionine	Tyrosine
Histidine	Aspartic acid
Arginine	Glutamic acid
Lysine	Proline
Leucine	Hydroxyproline

In addition, some of the nitrogen can be supplied from nonprotein sources such as urea. Therefore, as long as ruminants get an ample supply of energy nutrients and protein, with possibly some non-protein nitrogen (NPN) it is not necessary that any particular amino acid be included in the diet.

Protein Quality

The quality of a protein refers to the relative proportions of essential amino acids it contains compared to the requirements of the animal for those amino acids. A protein of highest quality would contain each of the essential amino acids in at least the amounts needed to meet the requirements of the animal. A protein of lesser quality may contain all of the essential amino acids, but one or more of them is present in amounts below what is needed. Some very poor quality proteins may not contain some of the essential amino acids at all.

Generally speaking, proteins from animal sources are of the highest quality when compared to monogastric animal requirements.

On the other hand, plant proteins are usually of lower quality. One exception is the protein of soybeans. Swine rations can be balanced for protein and energy using only corn and soybean oil meal. The quality of the protein from corn and the other cereal grains is generally low. If zein, the protein from corn, were used as the sole protein for swine, normal growth would not result. Corn protein is quite deficient in the essential amino acids lysine and tryptophan.

VITAMINS AND MINERALS

Vitamins

The vitamins are organic compounds that are required for normal growth and maintenance of the animal body. They are usually required in rather small amounts compared to carbohydrates and proteins. Vitamins are categorized by their solubility into the fat-soluble and water-soluble types. The fat-soluble group includes vitamins A, D, E, and K. The water-soluble group includes the B-complex vitamins and vitamin C.

Among the fat-soluble vitamins, vitamin A is the one most likely to be found deficient under present animal feeding schemes. This vitamin is necessary in the visual process; if it is not present in sufficient amounts night blindness may result. It also is necessary for the normal health of epithelial tissue of the reproductive, digestive, and respiratory tracts. Vitamin D is necessary in the normal metabolism of calcium and phosphorus in the bone. In most animals, the function of vitamin E is not clear. It has been demonstrated in rats to be necessary for normal reproductive function. Vitamin K is necessary for the normal clotting of blood.

Vitamin C is included among the water-soluble vitamins but it is not required in the diets of most domestic animals. The guinea pig is one of the few nonprimates that is unable to synthesize this vitamin at a rate rapid enough to meet its metabolic requirements. Humans and other primates, however, must have vitamin C in their diets. The several vitamins making up the B-complex group comprise most of the remaining water-soluble group. Most of the B-vitamins function in the utilization of fats, carbohydrates, and protein by the tissues. In this group are thiamine, riboflavin, niacin, pantothenic acid, and vitamins B_6 and B_{12}. Deficiencies of most of these vitamins are usually reflected in problems with the nervous system and epithelial tissue such as the skin. Some of the general functions and deficiency systems of certain vitamins are shown in Table 11-3.

TABLE 11-3. GENERAL FUNCTIONS AND DEFICIENCY SYMPTOMS OF SOME COMMON VITAMINS IN DOMESTIC ANIMALS

Vitamin	Functions	Deficiency Symptoms
FAT-SOLUBLE VITAMINS		
Vitamin A (and carotene)	Normal health of epithelium, visual cycle	Night blindness; skin and hair disorders; reduced growth and appetite
Vitamin D	Bone metabolism	Rickets; osteomalacia
Vitamin E	Reproduction in laboratory animals; antioxidant	Infertility in laboratory animals; muscle disorders
Vitamin K	Blood coagulation	Internal hemorrhage
B-COMPLEX VITAMINS		
Thiamine (B_1)	Carbohydrate, metabolism	Beriberi in humans; polyneuritis in birds; hyperirritability
Riboflavin (B_2)	Carbohydrate, protein, and oxygen metabolism	Curled-toe paralysis in chicks; impairments of the skin and nervous system
Niacin	Carbohydrate metabolism	Pellagra in humans; black tongue in dogs; dermatitis
Pantothenic acid	Component of enzyme system; important in several metabolic reactions	Goose-stepping in pigs; disorders of the nervous system, skin, and internal epithelium
Pyridoxine (B_6)	Tryptophan and protein metabolism	Dermatitis; convulsions; anemia in young animals
Cyancobalamin (B_{12})	Carbohydrate, fat, and nucleic acid metabolism	Pernicious anemia in humans; retarded growth; incoordination

TABLE 11-4. GENERAL FUNCTIONS
AND DEFICIENCY SYMPTOMS
OF SOME COMMON MINERALS
IN DOMESTIC ANIMALS

Mineral	Functions	Deficiency Symptoms
MACROELEMENTS		
Calcium (Ca)	Formation of normal bones and teeth; blood coagulation; muscle contraction	Rickets; osteomalacia
Phosphorus (P)	Formation of normal bones and teeth; numerous enzyme systems in metabolism	Retarded growth and poor appetite
Magnesium (Mg)	Enzyme systems	Muscle tetany; hyper-irritability
Potassium (K)	Muscle contraction; mineral balance	Weight loss; poor appetite
Sodium (Na)	Muscle contraction; osmotic pressure	Weight loss; poor appetite
Chlorine (Cl)	Osmotic pressure; digestion in the stomach	Poor appetite
MICROELEMENTS (TRACE ELEMENTS)		
Cobalt (Co)	Component of vitamin B_{12}	Retarded growth; anemia
Copper (Cu)	Enzyme systems; heart function	Retarded growth; poor appetite; anemia
Iodine (I)	Component of thyroxine	Goiter; low metabolic rate
Iron (Fe)	Component of hemoglobin; energy metabolism	Anemia
Manganese (Mn)	Enzyme systems	Slipped tendon in poultry; poor hatchability; lameness
Zinc (Zn)	Enzyme systems	Skin disorders in pigs; poor hatchability and feathering

Minerals

In contrast to the other nutrients, minerals are not compounds but individual chemical elements. Like the vitamins they are generally required only in small amounts for normal body function. Minerals are usually discussed in two groups, macroelements and micro-elements, according to the level at which they are required. The macroelements, even though they are required in smaller amounts than the proteins and carbohydrates, are required in larger amounts than the microelements or trace elements.

The macroelements include calcium, phosphorus, sodium, chlorine, potassium, sulfur, and magnesium. Calcium and phosphorus are required for normal bone and teeth formation. The four elements calcium, magnesium, potassium, and sodium are necessary for normal muscle contraction. Sodium and chlorine help regulate the osmotic pressure of body fluids. Chlorine is a component of hydrochloric acid produced in the stomach. Magnesium and phosphorus function in the metabolism of fats and carbohydrates. Sulfur is necessary in the synthesis of the sulfur-containing amino acids in ruminants.

Several elements are required in such small amounts that they are referred to as microelements or trace elements. Some of these elements are necessary components of other vital substances. Not all of the functions of each mineral has been presented here, but it should be clear from what has been presented that the minerals are certainly important in nutrition even though they may not be required in very large amounts. Some general functions and deficiency symptoms for several mineral elements are presented in Table 11-4.

STUDY QUESTIONS AND EXERCISES

1. Define each of the following terms:

carbohydrate	nutrient
cellulose	nutrition
disaccharide	osteomalacia
essential amino acid	rickets
fat	polysaccharide
globule	protein
glucose	protein quality
glycerol	starch
lipid	sugar
monosaccharide	vitamin
NPN	volatile fatty acid

2. Give some examples of nutrients that do not require digestion.

3. Why is digestion necessary?

4. Diagram a simplified reaction for photosynthesis.

5. Compare the respiratory processes of plants and animals in terms of inspiratory and expiratory products.

6. Name some carbohydrates that are found in animal tissues.

7. Compare the composition and natural sources of the three disaccharides—maltose, sucrose, and lactose.

8. Compare the relative sweetness of the six common sugars.

9. Explain how ruminants can obtain energy from cellulose.

10. What is the primary nutritional function of fats and carbohydrates?

11. Explain how fats can provide more energy than carbohydrates.

12. What are the three most common volative fatty acids produced in the rumen?

13. Compare the degree of saturation of plant fats to that of animal fats.

14. What functions do proteins serve in an animal?

15. What is meant by saying that a certain amino acid is essential?

16. List the essential amino acids for swine and rats?

17. Which amino acids are essential in human diets?

18. Why is there no list of essential amino acids for cattle and sheep?

19. What is meant by protein quality.

20. Which foods contain the highest-quality proteins for monogastric animals in general?

21. List the fat-soluble vitamins and indicate at least one function of each.

22. Name some vitamins that belong to the B-complex group.

23. In general, how do B-vitamins function?

24. List several mineral elements that function as components of vital organic compounds in the body.

25. Name several mineral elements that function in the metabolism of other nutrients.

26. Distinguish between the macro- and microelements.

12

Anatomy and Physiology of the Digestive System

The digestive systems of animals are in a large part adapted to the type of foodstuffs that each species generally consumes. Some animals consume other animals or animal products and are called carnivores. Their food is rather highly concentrated, so their digestive systems need not be as large as those of some other animals. Dogs and cats and the many species of wild animals related to them are primarily carnivorous, although they can and do consume varying amounts of plant materials. Other animals have digestive tracts adapted to the consumption of large amounts of fibrous plant materials and are called herbivores. Two large categories of herbivorous animals exist, those that have a stomach with a single compartment and those that have a stomach with several compartments. Animals possessing a single-compartmented stomach are called monogastric, a category which includes rabbits, guinea pigs, and horses. The carnivores are also monogastric animals. The ruminant animals such as cattle, sheep, goats, and their relatives are polygastric, having a stomach with four compartments.

Another group of mammals including swine, primates, rats, and mice are not as selective with their source of food and will consume both plant and animal material routinely. They are called omnivores. Poultry are generally omnivorous, although they are not equipped to handle very much of the fibrous parts of plants. Their digestive systems are relatively small.

Figures 12-1 and 12-2 show the digestive tracts of four domestic animals. The digestive system consists basically of a tube extending from the mouth to the anus or cloaca. It functions in the ingestion, mastication, digestion, and absorption of food and in the elimination of solid wastes.

ORGANS OF THE DIGESTIVE SYSTEM

The Mouth and Esophagus

Prehension includes those activities involved in getting food into the mouth. In domestic animals, the lips, teeth, tongue, beak, and front paws may be used in prehension. The lips of the horse are used extensively for prehension, as is the tongue of sheep and cattle. Hogs use an interesting method of prehension; they grasp food material with the teeth and then as they release their grasp on it, they quickly thrust their head toward the parcel of food causing it to move further into the mouth. Many of us can remember being reprimanded as children for "eating like a hog." Humans are more sophisticated in the matter of prehension in that we use forks and spoons and occasionally our fingers.

After the food is in the mouth, a certain amount of grinding or mastication may take place. In mammals, the teeth are the primary organs of mastication. Carnivores usually chew their food very little but use their teeth and front paws to tear their food into swallowing-size parcels. Herbivorous animals chew their food more to make it easier to digest when it reaches the stomach and small intestine. As the food is being chewed, it is mixed with saliva and formed into a bolus for easier swallowing. The food bolus is moved to the back of the mouth and into the pharynx with the aid of the tongue where the swallowing reflex is initiated. The esophagus is the swallowing organ. The bolus is carried down the esophagus by a wave of muscular contractions called peristalsis. In mammals, the esophagus leads directly into the stomach.

In poultry, the esophagus carries the food into the crop for temporary storage and then into the small stomach and gizzard.

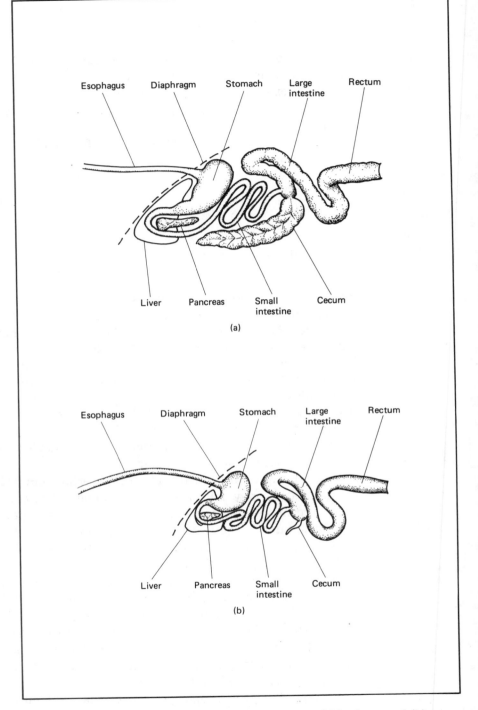

Figure 12-1   Diagram of the digestive systems of (a) a horse and (b) a pig.

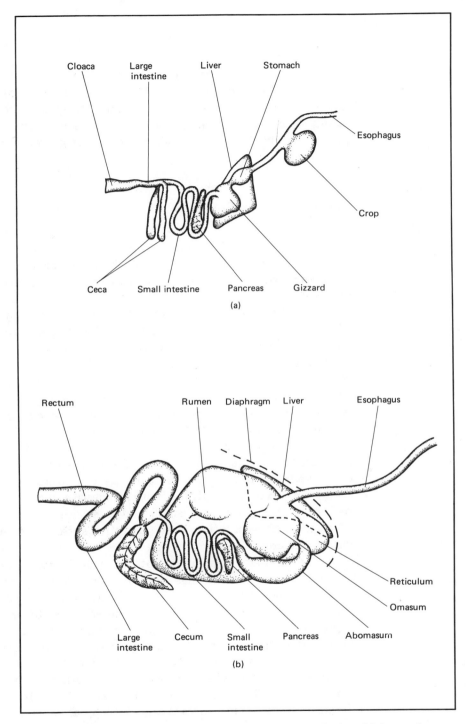

Figure 12-2 Diagram of the digestive systems of (a) a chicken and (b) a cow.

In birds the gizzard is the organ of mastication. The gizzard, or ventriculus, is a strong, muscular organ that functions in grinding and crushing food materials. The muscles in a turkey gizzard for example, are strong enough to crush a hickory nut. The grinding process is aided by the presence of grit or gravel taken in through the mouth.

The Nonruminant Stomach

The stomach is located just behind the left side of the diaphragm in monogastric mammals. It is a bean-shaped or crescent-shaped organ with the esophagus being connected in the concave side of the crescent.

One function of the stomach in monogastric systems is that of storage. A full stomach in a horse or pig may take a full day to empty provided another meal is not consumed before it empties. If it were not for the stomach, an animal could eat only small amounts at a time, and would therefore have to eat continually. The capacity of the stomach of a mature pig is approximately 8 ℓ, which is about 30 percent of the total capacity of the digestive system. The stomach of the horse has about an 18-ℓ capacity, which is less than 10 percent of the total capacity of the digestive tract. In dogs, the stomach makes up over one-half of the capacity of the digestive tract. In humans, the capacity of the stomach accounts for about one-sixth that of the entire digestive system. The stomach of birds is relatively very small and does not function as an organ of storage. This function in the digestive systems of birds is carried out by the crop.

The stomach is an organ of digestion in monogastric systems. The importance of the digestion that occurs in the stomach, however, is minor compared to that occurring in the small intestine. The only nutrient acted upon in the stomach to any appreciable degree is protein, and even then only partial digestion occurs. The protein-digesting enzyme called pepsin is secreted in the wall of the stomach by the chief cells. The enzyme requires activation by hydrochloric acid which is secreted by other gastric cells called parietal cells. The acid in the stomach also plays a role in the body's defense system against disease organisms by acting as an antiseptic.

The Ruminant Stomach

The stomach in ruminant animals such as cattle, sheep, and goats is made up of four compartments: The reticulum, the rumen, the omasum, and the abomasum. The total volume of the entire stomach makes up about two-thirds of the capacity of the entire digestive system in ruminant animals.

The rumen is the largest of the four compartments of the stomach in the polygastric system. The capacity of the rumen in a mature cow is about 200 ℓ, or over 50 gal. The rumen is located behind the left side of the diaphragm and occupies over one-half of the abdominal cavity. The rumen in a newborn calf makes up only about 30 percent of the total volume of the stomach. By the time the calf reaches a month of age the rumen has grown to about one-half the total volume of the stomach. In the two-month-old calf it represents about 70 percent of the gastric volume. In the mature cow the rumen represents about 80 percent of the total stomach volume. Not only does the rumen serve the obvious function of a storage organ, it also serves as a huge fermentation vat. Billions and billions of microorganisms, including hundreds of species of bacteria as well as some protozoa, live in the rumen. The microorganisms aid in the digestion of nutrients for the host. The relationship between the microorganisms and the host is called symbiosis, a process where both the host and the microscopic residents of the rumen benefit.

Rumination is a process that allows ruminant animals to graze and ingest food rapidly and chew it at their leisure at a later time. When grazing, cattle and other ruminant animals chew their food just enough to form a bolus and then swallow it. Later, the rather bulky boluses are regurgitated and chewed thoroughly, reinsalivated, and reswallowed. The rather finely chewed material disperses into the rumen contents and another bolus is regurgitated for further chewing. A cow may spend as much as eight hours per day ruminating, or chewing her cud.

The reticulum is located behind the diaphragm, adjacent to the heart, which is in front of the diaphragm. The nearness of the reticulum to the heart sometimes results in problems when sharp objects such as nails or wire are swallowed and fall into the reticulum. They may puncture the walls of the reticulum and diaphragm and penetrate the heart. The reticulum is sometimes called the "hardware" stomach. It is the smallest of the four compartments.

The omasum is the third compartment of the stomach in ruminants. The inside of the omasum contains many curtainlike leaves of tissue that provide a very large surface area. It is thought that this organ may play a role in squeezing and absorbing some of the excess fluids from the material passing from the rumen to the abomasum. The esophageal groove is formed by two heavy muscular folds of tissue that lead from the opening of the esophagus to the opening of the omasum. In young nursing calves this groove closes to allow milk to pass directly into the omasum and abomasum rather than into the rumen. When calves are bucket-fed with their heads down, the groove does not close and milk passes into the undeveloped rumen. This probably accounts for the paunchiness and digestive problems associated with bucket-fed calves. The use of nipple buckets usually corrects this problem because it causes the calf to raise its head and stretch its neck which apparently initiates the closure of the esophageal groove.

The abomasum is the fourth compartment of the ruminant stomach. The relative capacity of the abomasum in a mature cow is not much different than that of the stomach of a mature horse. The volume of the fourth compartment of the ruminant stomach makes up about 7 percent of the entire stomach. Enzymatic digestion in the abomasum is very much the same as that which occurs in the simple stomach.

The Intestines

The small intestine is the primary organ of digestion in animals. In ruminants a large amount of digestion occurs in the rumen, but this should not make us think that digestion in the small intestine of ruminants is any less important. Much of the digestion that occurs in the rumen is for the purpose of providing energy and protein for the microorganisms themselves. The microorganisms later become food for the ruminant animal and most undergo digestion in the small intestine. The small intestine is divided into three parts: the duodenum, the jejunum, and the ileum (not to be confused with ilium, a bone of the pelvis). The first part, the duodenum, constitutes about 5 percent of the length of the organ depending upon the species. A duct system leading from the pancreas and liver enters the first part of the duodenum.

The small intestine leads to the large intestine which likewise is made up of three parts: the cecum, the colon, and the rectum. The rectum terminates with the external opening called the anus in mammals or the cloaca in birds. The size of the cecum varies tremendously from species to species. In carnivores (dogs and cats) and humans, its capacity is only a few ounces. In sheep and pigs it may have a capacity of 1 or 2 ℓ. In cattle, the volume of the cecum may be 5 to 10 ℓ. In herbivorous monogastric animals, the capacity of the cecum is relatively large. In horses this may be from 20 to 30 ℓ. The function of the cecum in horses, rabbits, and guinea pigs is similar to that of the rumen in cattle: It helps carry out the digestion of fibrous foodstuffs through the aid of micro-organisms. A major difference is the fact that the cecum is located near the end of the tract while the rumen is near the beginning.

The large intestines of dogs and poultry are the smallest of the domestic animals in comparison to the animal's overall size. That of the horse is the largest and most complex.

The Accessory Organs

Some glands and organs that are not actually a part of the digestive tract play an important role in the process of digestion. These include the liver, pancreas, and salivary glands. The salivary glands secrete a fluid called saliva which helps to lubricate the food for easier swallowing. In some species, the saliva contains a small amount of enzyme that aids in the digestion of starches. The liver secretes bile which is stored in the gall bladder. Bile plays a part in the digestion of fats. The pancreas secretes several digestive enzymes which are emptied into the duodenum of the small intestine.

DIGESTION AND ABSORPTION

The function of digestion is to change large complex nutrients such as starch and protein into small easy-to-absorb nutrients such as sugars and amino acids. Even if the larger complex materials were to be absorbed, they could not be utilized at the cell level. Carbohydrates that require digestion to be utilized include cellulose, starch, glycogen, and the disaccharides. Proteins, peptides, fats, and nucleic acids also need to be digested. Vitamins and minerals do not require digestion. They are usually absorbed and utilized in the forms in which they occur in the food.

Digestion of Carbohydrates

In most monogastric animals, the digestion of carbohydrates begins in the small intestine. Starches that are carried into the small intestine from the stomach come in contact with an enzyme called amylase. One type of starch is called amylose; the enzyme that functions in its digestion is called amylase. The -*ase* ending indicates that it is an enzyme. The prefix usually indicates the substance being digested. Amylase is secreted by the pancreas and carried into the duodenum. As starch is being digested, fragments called dextrins are produced. These are eventually also broken down to maltose. Digestion is not complete, however, until maltose is broken down to glucose. The enzyme that helps carry out this process is called maltase. It is secreted by the small intestine.

Once the free glucose is produced, it is absorbed through the walls of the small intestine into the blood stream. The following diagram illustrates the complete process of digestion of starch in the small intestine:

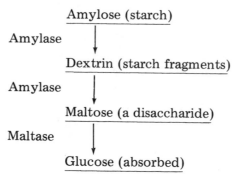

Other disaccharides such as sucrose and lactose may be consumed also. Both of these require digestion before they can be absorbed and utilized. Even if disaccharides were to find their way into the blood stream, they could not be utilized by the body cells. Sucrose is digested by the intestinal enzyme sucrase to yield the monosaccharides glucose and fructose. Lactose can be digested by the enzyme lactase to yield glucose and galactose. Both galactose and fructose can be utilized by body cells since they are monosaccharides. However, both are usually converted to glucose by the liver before they have a chance to be utilized to any great extent.

Cellulose is a more complex carbohydrate than starch and cannot be digested in the small intestine. No enzymes are secreted by mammals and birds that are capable of breaking it down. However, microorganisms do secrete such enzymes. Therefore, in the rumen of ruminant animals and in the ceca of most animals, microbial digestion of cellulose does occur. The complete digestion of cellulose ultimately produces glucose. Most of this glucose is absorbed by the microorganisms themselves to be used for energy purposes. In the process of metabolizing glucose, the microorganisms produce by-products called volatile fatty acids (VFA). These acids are readily absorbed through the rumen wall and dissolved in the blood. The blood carries the VFA to the cells where they can be utilized for energy. The presence of the microorganisms thus enables ruminants to utilize an otherwise useless feedstuff. It is quite possible for some ruminants to produce enough VFA in the rumen to provide their total daily requirement of energy. A similar process occurs in the cecum, but not to the extent that it occurs in the rumen.

It should be pointed out that starches are also digested by the microorganisms in the rumen with much the same outcome as from cellulose. Not all of the starches and cellulose undergo complete digestion while in the rumen. This means that along with the microorganisms that spill over into the abomasum and small intestine go some undigested and incompletely digested carbohydrates. The starches, dextrins, and disaccharides will undergo further digestion. Any undigested cellulose may undergo further digestion in the cecum or be carried out in the feces. Examination of the feces of herbivorous animals will show certain amounts of cellulose.

Digestion of Proteins

Protein digestion in monogastric animals begins in the stomach under the influence of gastric protease. The older name for this enzyme is pepsin. Pepsin is secreted in an inactive form and must be activated by hydrochloric acid before it can properly function. Protein digestion does not proceed to completion in the stomach because food does not remain in the stomach for a long enough period of time. As protein moves into the small intestine, it is exposed to a number of enzymes that may carry digestion to completion. As with fats and carbohydrates, the digestion of proteins is not 100 percent efficient and some of each of these nutrients will pass through the digestive system and into the feces. Two proteases produced in the pancreas and directed to the duodenum are trypsin and chymotrypsin. The effects of the two pancreatic proteases and gastric protease are very similar in that the protein molecule is split up into fragments called peptides and polypeptides with some free amino acids being produced. Other enzymes called peptidases help in the completion of protein digestion. The following diagram illustrates the complete digestion of protein:

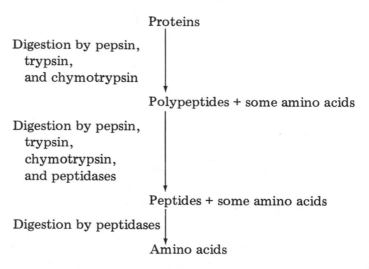

As free amino acids are produced, they are absorbed into the blood to be carried to tissues and cells where they will be used to synthesize various proteins or metabolized for other purposes.

In the rumen of cattle, sheep, and goats, proteins are first exposed to the action of proteases produced by the microorganisms. These organisms convert proteins to amino acids and use the amino acids to synthesize proteins within their own cells. If certain amino acids are not available in the proteins consumed by the host, the microorganisms will synthesize them from the available protein or from nonprotein nitrogen and other carbon sources. The protein that is ultimately synthesized by the microorganisms is a rather high-quality protein, nearly comparable to that of meat, milk, and eggs. The microbial protein passes from the rumen on into the abomasum and small intestine in the form of living and dead cells of bacteria and protozoa. Some undigested feed proteins from the rumen may also find their way down the digestive tract. All of this protein is acted upon by pepsin from the abomasum and by the proteases and peptidases found in the small intestine. Protein digestion in the abomasum and small intestines of ruminants is generally the same as that which occurs in mongastric systems. Some microbial digestion also may occur in the cecum of herbivorous animals whether ruminant or monogastric.

Digestion of Fats

An enzyme called lipase, which is secreted by the pancreas and emptied into the duodenum of the small intestine, is the chemical force behind the digestion of fats. Because of the tendency of fat molecules to cling together in globules, the digestion of fats might be expected to proceed very slowly. A substance called bile, which is secreted by the liver and stored in the gall bladder, also has an important role to play in fat digestion. Bile is a detergent, or soap. When it comes in contact with fat globules it causes the molecules to disperse. This dispersion of the fat molecules allows the pancreatic lipase to react with many more molecules of fat in a much shorter period of time. The presence of bile thus enables fat digestion to proceed rather quickly. The digestion of fats results in the production of glycerol and fatty acids which can then be absorbed. The breakdown of the fat globule by bile is called emulsification.

STUDY QUESTIONS AND EXERCISES

1. Define each of the following terms:

abomasum	amylase
absorption	bile

bolus   mastication
carnivore   monogastric
cecum   omasum
chlorophyll   omnivore
cloaca   pepsin
colon   peristalsis
cud   polygastric
dextrin   prehension
digestion   rectum
duodenum   regurgitation
esophagus   reticulum
gastric   rumen
herbivore   ruminant
ileum   rumination
ingestion   symbiosis
intestine   trypsin
jejunum   ventriculus
lipase

2. Compare the way in which mammals and birds store food temporarily in their digestive systems.

3. Why is temporary storage of food in the digestive tract important?

4. Compare the process of mastication in birds and mammals.

5. Compare the relative capacities of the stomachs of domestic animals.

6. Compare the importance of the stomach and small intestine as digestive organs in monogastric systems.

7. What is the function of hydrochloric acid in the stomach?

8. Compare the relative volumes of the four compartments in the mature ruminant stomach.

9. Compare the stomach of a newborn calf to that of a mature cow.

10. How does the rumen differ from the abomasum and small intestine as an organ of digestion?

11. Explain the function of the esophageal groove in young ruminants.

12. Compare the size and importance of the cecum in the various domestic animals.

13. Compare the importance of the cecum of a horse to that of the rumen in a cow.

14. Name the accessory organs of digestion and explain why they are called accessory.

15. Compare the digestion of starch and cellulose.

16. Name at least three digestive enzymes secreted by the pancreas.

17. Explain the process of protein digestion in the simple stomach and small intestine.

13

Feeds and Feeding

The importance of knowing about the chemistry of nutrients and the physiology of animal digestive systems becomes apparent when livestock feeders must select and provide the proper combinations of feeds to meet the needs of their animals. In order to choose the proper feeds, the feeder must know what nutrients those feeds contain and whether they will be available to the animals. Feeds must be analyzed for nutrient content and digestibility. The nutrient requirements for various kinds of body activities must be known. Finally, the ration or diet of the animals must be balanced with the most economical of the available feeds. This is what feeds and feeding is about.

CLASSIFICATION OF FEEDSTUFFS

The term feedstuff means the same as feed or food but also includes some materials that often may not be thought of as feed. An example of such a material would be a refined amino acid or synthetic vitamin mixture. Feedstuffs are categorized into three groups based upon their fiber content and nutritional use: roughages, energy feeds, and protein supplements. Although some feed additives do not supply nutrients to an animal, they will be discussed along with the feedstuffs.

Roughages

The primary characteristic of roughage that distinguishes it from the other categories of feedstuffs is its fiber content. Much of the fibrous parts of plants is made up of cellulose. Other substances found in plant fiber are lignin and a celluloselike material called hemicellulose. As we shall see, fiber is usually less digestible than other plant materials such as starch and protein. On a dry weight basis, roughages may contain anywhere between 25 and 40 percent fiber. Fibrous feedstuffs are bulky—they weigh less than more concentrated feeds for the volume they occupy. Most of the common roughages are either grasses or legumes. Grasses are plants having bladelike leaves such as wheat, orchardgrass, and corn. Grasses produce a kind of seed referred to by the term monocotyledon. Legumes are dicotyledonous and have a more rounded leaf. From a nutritional point of view, legumes contain more protein. Examples of legumes are the clovers, beans, peas, and alfalfa.

Roughages may be further classified according to the method by which they are harvested and fed. They may be succulent or dry. Succulent roughages include pasture, silage, and greenchop forage. The term forage can be used for roughage. Pasture is probably the most economical source of forage because the grazing animal does the harvesting. The term meadow can be used to refer to pastureland. However, pasture usually refers to a field used for grazing purposes only, whereas a meadow can be grazed or harvested for hay. As is true with any roughage, the nutrient quality of pasture is affected by the level of fertility in the soil in which it grows. For example, the application of nitrogen fertilizers will result in increased protein content of grass pastures. Soil tests should be taken and applications of fertilizer made to obtain the optimum quality and quantity of forage.

Greenchop forage is that which is harvested and fed to livestock immediately. This is sometimes done when animals are being kept in confinement. As is true with pasture and other succulent feeds, the water content of greenchop forage is rather high, possibly 60 to 80 percent. Silage is another succulent roughage. It is forage which has been stored in a relatively airtight structure called a silo. Anaerobic microorganisms in the forage produce a substance that prevents it from spoiling. Forage can sometimes be harvested for silage when it cannot be harvested for hay because of weather and other conditions.

The most common form of dry roughage is hay. The plant is cut and dried in the field until it has less than about 25 percent water. It is then removed from the field and stored in a dry building or stacked and covered. Hay may be baled or stored loose, although in recent years most has been stored as baled hay. Sun-cured or sun-dried hay usually has a certain amount of vitamin D due to the effect of sunlight on the ergosterol content of the dead plant. Overexposure to sunlight, however, can result in a considerable loss of other nutrients. Legume hays have a greater protein content than do the grass hays. Refer to Table 13-1 and compare the protein content of alfalfa, lespedeza, and the clover hays to that of brome, fescue, and orchardgrass hays. The stage of maturity of the crop when harvested has an effect upon the nutrient content of the hay as well as the yield. The younger the crop, the greater the percentage of protein and water, and the lower the fiber content. However, the yield will also be less. As the plants mature they become larger and taller and thus the yield in tons of hay per acre increases. At the same time, the protein content decreases and the fiber content increases. For example, the fiber content of orchardgrass hay as a percentage of the dry matter content is approximately 28, 33, and 39 percent for hay cut in the immature, full bloom, and mature stages, respectively. The protein content of the same three stages are about 17, 12, and 6 percent, respectively. The problem becomes one of obtaining the optimum yield and quality of hay for the particular kind of animal to be fed. Other examples of dry roughage include straw, stover, and fodder. Straw usually refers to the stemmy parts of plants after the seeds have been removed, such as wheat or oat straw. Stover is similar to straw but usually refers to plants such as corn or sorghum. Fodder is the dry, cured stalks and leaves of corn and the sorghums including the grain.

TABLE 13-1. APPROXIMATE CONTENT OF SOME COMMON FEEDS FOR CATTLE AND SWINE

Feed	Pounds of Dry Matter per Pound of Feed	Pounds of Total Protein per Pound of Feed	Pounds of Digestible Protein per Pound of Feed		Pounds of TDN per Pound of Feed		Megacalories of Digestible Energy (DE)	
			Cattle	Swine	Cattle	Swine	Cattle	Swine
Energy Feeds (Grains)								
Barley	0.89	0.12	0.09	0.09	0.72	0.71	1.44	1.42
Corn	0.86	0.09	0.065	0.07	0.78	0.79	1.57	1.59
Oats	0.89	0.12	0.087	0.095	0.66	0.62	1.33	1.24
Sorghum milo	0.89	0.09	0.05	0.06	0.71	0.79	1.43	1.58
Wheat	0.89	0.12	0.09	0.105	0.78	0.78	1.57	1.64
Protein Feeds								
Alfalfa meal (dehydrated)	0.92	0.18	0.14	0.08	0.57	0.33	1.14	0.67
Cottonseed meal	0.91	0.42	0.34	—	0.69	—	1.49	—
Meat & bone meal	0.94	0.50	0.45	0.44	0.61	0.82	1.21	0.90
Soybean meal	0.89	0.46	0.39	0.42	0.72	0.75	1.45	1.50
Tankage	0.94	0.59	0.54	0.38	0.66	0.68	1.31	1.36
Roughages								
Alfalfa hay	0.91	0.16	0.10	0.70	0.49	0.32	0.99	0.65
Bromegrass hay	0.90	0.11	0.06	—	0.50	—	1.00	—
Tall fescue hay	0.87	0.08	0.045	—	0.53	—	1.05	—
Lespedeza hay (Korean)	0.92	0.13	0.08	—	0.53	—	1.06	—
Orchardgrass hay	0.89	0.10	0.06	—	0.53	—	1.06	—
Red clover hay	0.90	0.13	0.08	—	0.53	—	1.05	—
Corn silage	0.28	0.02	0.014	—	0.20	—	0.40	—
Sorghum silage	0.31	0.03	0.013	—	0.18	—	0.35	—

Source: Adapted from National Research Council data.

Energy Feeds

The function of the second group of feedstuffs is to meet the energy or caloric needs of animals. This group includes the cereal grains— corn, oats, barley, wheat, rye, and rice. It also includes any other feedstuff with a high starch content such as root crops, fruits, and nuts. Energy feeds are sometimes called basal feeds since energy is the single largest nutritional requirement among animals. Energy itself is actually not a nutrient, although it is provided by such nutrients as starch and fat. Another term that is often used to refer to energy-producing feeds is concentrates, or energy concentrates. This term suggests a high concentration of energy-producing nutrients, usually starch. The amount of starch in the feed grains is indicated by the nitrogen-free extract (NFE) portion of a feed analysis. This terminology will be explained later in this chapter. Table 13-2 presents the NFE content of several feed grains as well as other components as a percentage of the dry matter content of the feed.

TABLE 13-2. COMPONENTS OF SELECTED FEEDS AS A PERCENTAGE OF DRY MATTER

	NFE	Fiber	Total Protein	Fat
Barley	77%	5.6%	13.0%	2.1%
Corn	81	2.3	10.6	4.9
Oats	66	12.4	13.2	5.1
Sorghum milo	80	2.2	12.4	3.1

Energy feeds are usually relatively low in fiber and fat. Oats are somewhat of an exception in that the fiber content is quite a bit higher than that of corn or milo. This is due to the presence of the hull around the oat kernel. Notice, also, that oats have the lowest NFE or starch content of the four grains shown in Table 13-2. Feed grains as they are fed usually contain from 10 to 15 percent water. Feeds with this amount of moisture are usually called air-dry. Compared to roughages, energy feeds are not bulky, primarily because of their low fiber content. Oats, however, because of their higher fiber content, are more bulky than corn.

Protein Supplements

The nutrient required by animals in the second-largest amount is protein. If a group of animals is fed a type of feed that contains sufficient carbohydrate and fat to meet their energy needs, but the protein content of that feed is below their requirement, a supplement containing a higher percentage of protein must be provided. The term protein concentrate is often used to refer to protein feeds. Animals require from 10 to 20 percent protein in their diets depending on their age and type of activity. Any feed containing a protein content of about 20 percent or higher is usually referred to as a protein feed. Compare the total protein contents of energy feeds, protein feeds, and roughages in Table 13-1.

Protein feeds are obtained from a number of sources, including plants, mammals, birds, and marine animals. Many of the protein feeds commonly used are obtained as by-products of other industries. One such group of protein supplements is the oil meals. Fats used in the human food industry are obtained from soybeans, cottonseed, flaxseed, corn, and other seeds. The seeds of soybeans, cottonseed, and flaxseed contain a rather large amount of protein as well as fat. After the fat has been removed, the protein can be used for feed for domestic animals. Soybean protein is used in some human foods, also. Soybean oil meal and cottonseed meal are used extensively as protein supplements for domestic animals.

Another by-product protein supplement category includes the cereal grain by-products such as the germ and gluten of seeds. The part of the seed that contains the endosperm is called the germ. This is the part of the seed from which the small sprout will develop. The gluten is found in the seed capsule. This is not the same as the hull for those seeds that have hulls. Examples of protein feeds in this category include corn gluten meal, corn germ meal, wheat germ meal, and grain sorghum gluten meal. These products remain after the starch has been removed from the seed. Other by-product protein feeds obtained from cereal grains include those obtained from the alcohol distilling industries. Alcohol is obtained from the action of yeast on the starch in the grain. After the process is completed, the relative protein content of the grain by-product is considerably increased because much of the starch has been removed. Examples of such feeds are barley distillers' dried grains and wheat distillers' dried solubles. Protein feeds may be obtained from animal as well as plant by-products. Meat meal and digester tankage are obtained as by-products of the meat-packing industry. These protein feeds contain the rendered, dried, and ground meat and bone by-products of animals that have been slaughtered. Hair, horns, and hooves are not included. Dried blood meal is another example of an animal by-product with an exceptionally high protein content. Dried skim milk is yet another example of an animal by-product used for a protein feed.

Feed Additives

Most feed additives are nonnutrient compounds that are added to the livestock rations to promote efficiency of feed utilization. They may do this by promoting growth and production, by decreasing the effects of stress, or by reducing susceptability to disease. Some products such as synthetic vitamins and other refined materials are sometimes classified as feed additives but are also nutrients.

Stilbestrol, actually diethylstilbestrol (DES), has been used to promote rate of gain and feed efficiency in beef steers. The use of DES was recently banned by the Food and Drug Administration (FDA). Thyroprotein, or iodinated casein, has a thyroxinelike activity and has been fed to young animals to promote increased growth. Increases in milk production have been noted after feeding of iodinated casein to dairy cattle. It is not used for dairy cows on official production testing programs. Antibiotics are used extensively in poultry and swine rations to increase gains and efficiency. Rumensin is the trade name of another feed additive that will produce increases in rate of gain and efficiency in beef cattle.

Some growth stimulants are designed not to be added to the feed, but to be implanted beneath the skin on the ear. Ralgro and Synovex have been approved by the FDA for implanting in cattle, except for breeding cattle. These implants provide stimulation in rate and efficiency of gain similar to that obtained from DES implants.

The Proximate Analysis

In order to do an effective job of feeding animals, the feeder must know something about the nutritive value of the feed. The most common method for analyzing feedstuffs to determine their content of carbohydrates, fats, and protein has been the proximate analysis. This system of feed analysis was developed by two German scientists at the Weende Experiment Station in Germany in 1864, about the same time that Mendel was discovering some rather important principles of genetics. The proximate analysis separates feedstuffs into six fractions as shown in Table 13-3.

The percentage of water is determined by weighing a finely ground sample in an oven at 100°C until it attains a constant weight. The difference in the weights of the sample before and after the drying process represents the water content of the feed sample. The percent water is determined using the following formula:

$$\% \text{ water} = \frac{(\text{Weight of "wet" sample} - \text{weight of dried sample}) \times 100}{\text{Weight of "wet" sample}}$$

The percent dry matter can be determined simply by subtracting the percent water from 100.

Ash is determined by burning a feed sample at between 500° and 600°C. The weight of the ashed sample is expressed as a percent-

TABLE 13-3.   PROXIMATE ANALYSIS COMPONENTS

Fraction	Major Components
Water	Water and volatile compounds
Ash	Minerals
Ether extract	Fats, waxes, and some pigments
Crude fiber	Cellulose, hemicellulose, and lignin
Crude protein	Protein, amino acids, and other nitrogenous compounds
Nitrogen-free extract	Starch and sugar

age of the original sample weight. The content of ash is quite variable depending upon the feedstuff. It is of very little value in measuring a particular mineral element.

The ether-extract fraction measures fat, free fatty acids, cholesterol, chlorophyll, and certain other lipids. Most common feedstuffs generally contain only small amounts of fat. In addition, the proportion of nonfat lipids in most feeds is far less than the amount of fat present. This fraction is determined by mixing a finely ground feed sample in petroleum ether for a specified amount of time. The ether is then separated from the undissolved material and evaporated. The lipid residue is weighed and expressed as a percentage of the weight of the original sample.

Crude fiber is a measure of the amount of cellulose and much of the hemicellulose and lignin of a feedstuff. It is determined by boiling a finely ground sample first in a weak solution of sulfuric acid and then in a weak solution of sodium hydroxide. The weight of the undissolved residue after drying is expressed as a percentage of the original sample weight. The two boiling processes result in complete removal of protein, starch, and sugar. Cellulose is only slightly affected by the acid and alkali. Variable amounts of lignin and hemicellulose are dissolved depending upon the feedstuff.

The crude or total protein content of a feed sample is calculated by multiplying the nitrogen content times the constant 6.25. This figure is based upon the fact that protein contains approximately 16 percent nitrogen (16% × 6.25 = 100%). The amount of nitrogen is determined by a procedure called the Kjeldahl method. If a protein contains much more or less than 16 percent nitrogen, the figure 6.25 will be inaccurate. For example, the nitrogen content of wheat protein is generally higher than 16 percent so a factor of 5.7 is used. Milk protein usually contains less than 16 percent nitrogen, so a factor of 6.4 is often used when analyzing milk products.

The preceding five procedures involve actual laboratory analysis and account for most of the nutrients except the vitamins, sugars, and starches. Vitamin content is not determined by the proximate analysis. Sugar and starch content, however, can be estimated by subtracting the percentages of all the other five fractions from 100. The result is a measure of the sugars, starches, and small amounts of the more soluble hemicelluloses. It has been given the name nitrogen-free extract or NFE.

As an example, suppose that a sample of oats grain has been

analyzed by the proximate analysis method and produced these results: 11 percent water, 3.2 percent ash, 4.5 percent ether extract, 11 percent crude fiber, and 1.9 percent nitrogen. The crude protein is determined by multiplying the percent nitrogen by 6.25.

$$1.9\% \text{ nitrogen} \times 6.25 = 11.9\% \text{ crude protein}$$

The percent NFE is determined by using the following formula:

$$\% \text{ NFE} = 100\% - (\% \text{ water} + \% \text{ ash} + \% \text{ fiber} + \% \text{ ether extract} + \% \text{ protein})$$

When the actual values resulting from the proximate analyses are substituted we obtain

$$\% \text{ NFE} = 100\% - (11\% \text{ water} + 3.2\% \text{ ash} + 11\% \text{ fiber} + 4.5\% \text{ ether extract} + 11.9\% \text{ protein})$$
$$= 100\% - 41.6\%$$
$$= 58.4\%$$

Notice that in the formula the percent protein is used rather than the percent nitrogen.

A relatively new method of separating the dry matter portion of feeds into two fractions, cell contents and cell-wall substances has been suggested by Van Soest. The proposal is based upon the fact that the cell-wall contents are composed mainly of cellulose and lignin while the more digestible fats, proteins, sugars, and starches are found in the cell contents. This procedure is essentially another method for determining the crude fiber content of a feedstuff.

Digestibility and TDN

Knowing the approximate nutrient content of feedstuffs does not tell us how much of a particular nutrient is actually digested and metabolized by an animal. To determine what proportion of individual nutrients is digested and absorbed, digestion trials are conducted. Different species of animals will vary in their ability to digest the various nutrients. For this reason, digestion trials must be conducted with different classes of animals in order to determine digestibility coefficients for a particular feed when fed to a particular kind of animal. An animal participating in experiments will

be fed weighed quantities of a given feed for several days before
the actual trial begins. This is done so that the digestive system
will contain only the feed being tested. During the actual trial,
the same amount of feed is given to the animal each day. The feces
are collected and weighed and analyzed in the same manner as the
feed was analyzed. The amount of each nutrient found in the feces
represents that part that was not digested by the animal. The amount
that was digested and absorbed is determined by subtracting the
amount of each nutrient in the feces from the amount in the feed.

Table 13-4 contains the results of a digestion experiment.
In column 2 are listed the amounts of the various nutrients in 3.5 lb
of hay as determined by the proximate analysis. In column 3 are
listed the amounts of each nutrient expressed as a percent of the
3.5 lb of hay. The figures in column 4 are the amounts of the various
nutrients in 7.2 lb of feces as determined by a proximate analysis.
Notice that the water content of the feces is much higher than that
of the feed, about 83 percent. This accounts for the greater amount
of feces produced by the animal. The amounts of each nutrient
digested, as shown in column 5, is determined by subtracting the
amount in the feces in column 4 from the amount in the feed in
column 2. For example, 0.04 lb of fat (line 4) was found in the
feces when there was 0.09 lb originally in the feed. This means that
0.05 lb (0.09 lb – 0.04 lb = 0.05 lb) was actually digested and absorbed by the ewe. No data is presented in columns 5, 6, and 7
for water and ash. The amount of water consumed by the animal
is not taken into account and the amount of ash digested is mean-
ingless in calculating nutrient digestibility. Mineral matter and water
are absorbed directly and do not need to be digested. The figures
in column 6 indicate the amount of each of the digested nutrients
expressed as a percent of the original 3.5 lb of hay. For example,
the 0.5 lb of fiber that was digested (line 5, column 5), represents
14.3 percent of the original 3.5 lb of feed. The figures in column
7 are called digestion coefficients. They indicate the amount of
each nutrient that was actually digested. For example, of the 0.44 lb
of protein that was present in the hay, only 57 percent was actually
digested (0.25/0.44 × 100). Depending on the type of feed, coefficients of digestion may be expressed either in the decimal form or as
percentages (57% or 0.57).

Once the data in Table 13-4 has been determined, it is a relatively easy matter to calculate the percent TDN (total digestible

TABLE 13-4. RESULTS OF A DIGESTION TRIAL INVOLVING A 120-LB EWE BEING FED 3.5 LB OF RED CLOVER HAY PER DAY

Line Number	Item (1)	Amount of Hay Fed Pounds (2)	Amount of Hay Fed Percent (3)	Weight of Feces (lb) (4)	Amount Digested Pounds (5)	Amount Digested Percent (6)	Coefficient of Digestibility (%) (7)
1.	Feed and feces	3.50	100.0	7.20	—	—	—
2.	Water	0.41	11.7	5.97	—	—	—
3.	Ash	0.22	6.3	0.09	—	—	—
4.	Fat	0.09	2.6	0.04	0.05	1.4	56
5.	Fiber	0.95	27.1	0.45	0.50	14.3	53
6.	Nitrogen	0.07	2.0	0.03	0.04	1.1	—
7.	Protein (N × 6.25)	0.44	12.6	0.19	0.25	7.1	57
8.	Subtotal (sum of lines 2, 3, 4, 5, and 7)	2.11	60.3	6.74	—	—	—
9.	NFE (line 1 minus line 8)	1.39	39.7	0.46	0.93	26.6	67

nutrients). The TDN is a measure of the energy value of a feed. It is determined essentially by adding the figures for the four digested nutrient fractions of the proximate analysis. However, before this can be done, the value for fat must be multiplied by the factor 2.25. As we explained in the previous chapter, fat contains about 2.25 times more energy than do the carbohydrates when they are metabolized. For a feed such as red clover hay, the error would not be too serious if this procedure was omitted. However, for feeds that contain considerably higher amounts of fat, the error would become more serious. The following formula can be used to calculate TDN:

% TDN = (% digestible fat $\times$ 2.25) + % digestible fiber +
% digestible protein + % digestible NFE

Using the data from Table 13-4, the TDN content of red clover hay can be calculated as follows:

% TDN = (1.4% digestible fat $\times$ 2.25) + 14.3% digestible fiber +
7.1% digestible protein + 26.6% digestible NFE

= 51.2%

Caloric Measures of Feed Value

Energy is quite often the factor that limits livestock production because it is required by animals in the largest amounts compared to other nutrients. Much of the comparative evaluation of feedstuffs, then, is based upon their ability to supply energy to the animal. A method of measuring the energy value of a feed that is more direct than the proximate analysis is the caloric method. The total or gross energy in a feed can be determined by burning it in an instrument called the bomb calorimeter. A feed substance is burned inside a closed chamber filled with oxygen. The change in the temperature of a known amount of water surrounding the chamber is measured. The amount of heat given off by the substance when completely burned is equivalent to the total energy it contains. The energy units used to measure this heat are called calories. One calorie is defined as the amount of heat required to raise the temperature of 1 g of water from 15° to 16°C. Since this is a relatively small unit, the kilocalorie (1 kcal = 1000 cal) or megacalorie (1 Mcal

= 1000 kcal = 1,000,000 cal) is more often used. In writings associated with human nutrition, the term Calorie (with an uppercase C) is used in place of kilocalorie. Unfortunately this often leads to confusion. The magacalorie is sometimes called a therm. Another unit of heat measurement is the British thermal unit or BTU. One BTU is the amount of heat required to raise 1 lb of water from 38.7° to 39.7°F. In making conversions between the British and metric systems, it is helpful to remember that 1 BTU is equal to about 252 cal, and 1 kcal is equal to about 4 BTU.

The amount of gross energy varies with types of compounds. Carbohydrates contain slightly over 4 kcal per gram of substance. Proteins contain about 5.6 kcal/g and fats about 9.5 kcal/g. Different degrees of digestibility of two feeds has nothing to do with their gross energy values. For example, starch contains about 4.2 kcal/g while oat straw contains approximately 4.4 kcal/g.

Not all of the gross energy is useful to the animal. A measure of digestible energy (DE) can be determined by measuring the gross energy in the feces and subtracting it from the gross energy of the feed. A digestion trial must be conducted as explained for the proximate analysis of feed and feces. If digestible energy is expressed as a percent of gross energy, the resulting figure is very similar to the percent TDN. For example, the gross energy of red clover hay is approximately 4.4 kcal/g. The digestible energy is approximately 2.29 kcal/g. The percent digestible energy is 52 (2.29/4.4 × 100). Compare this figure to the percent TDN of 52.2 calculated in the previous section. The TDN and DE systems of feed evaluation are widely used as measures of a feed's energy value.

There are other losses of energy to the animal in addition to that lost in the feces. When these other losses are taken into account, more accurate means are obtained to determine the energy value of feeds. One of these is metabolizable energy. As shown in Figure 13-1, when digestible energy is partitioned and an accounting is made for that which is lost in the urine and combustible gases, the result is the energy that is actually available at the cell level for metabolism. The gases are particularly important in calculating metabolizable energy (ME) in ruminants. Considerable amounts of methane gas are produced in the rumen and lost through the mouth—as much as 6 to 9 percent of the gross energy consumed. Since the methane is contained in the expired air, it is rather difficult to measure. Extensive studies in ruminants have shown that the ratio of DE to ME is rather constant. Therefore, it is quite common to estimate ME in ruminants by multiplying the DE by 0.82. In poultry, ME is easier to obtain than DE due to the fact that the urine and feces are expelled through a common opening. In nonruminants the amount of combustible gases produced in the digestive system are usually negligible so that ME is usually determined by considering only the fecal and urinary losses.

Net energy values are determined by subtracting the heat generated from nutrient metabolism and rumen fermentation from metabolizable energy. There is also a considerable amount of fermentation occurring in the ceca of herbivores which lends to this loss of energy which has come to be known as the heat increment. Net energy (NE) is what remains of the gross energy (GE) in the feed after all losses have been taken into account.

$$NE = GE - (FE + UE + CG + HI)$$

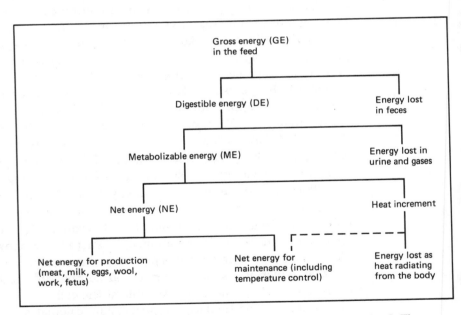

Figure 13-1   Partition of feed energy as it is used by an animal. The dashed line indicates the heat generated as a result of the work of digestion and absorption which is used to help maintain body temperature when the environmental temperatures are relatively low.

In this formula, FE is fecal energy, UE is urinary energy, CG is combustible gases, and HI is the heat increment. During times when the environmental temperature is relatively low, part of the energy generated by body metabolism is used to help maintain body temperature, so that part is not actually a loss. Net energy can be further subdivided into the net energy needed to maintain body weight and condition of the animal, and that needed for the animal to produce useful materials and work. The terms NE_m and NE_p are used to represent the net energy of maintenance and the net energy of production, respectively. As shown in Figure 13-1, net energy for production, includes that needed for growth, fattening, milk production, egg production, wool growth, fetal development, and other useful work.

ANIMAL FEEDING

Requirements for Nutrients

Nutrients obtained from feeds are necessary to support such body functions as maintenance, growth, lactation, reproduction, and work. The first requirement of an animal is to consume enough of the various nutrients to keep basic body functions going at such a rate as to not gain or lose weight. Basic metabolic functions of cells must be maintained. An animal that is not growing or producing useful products needs energy to maintain respiratory, kidney, and circulatory functions. Minimal amounts of protein are needed to replace cells that wear out during basal metabolism. Vitamins and minerals must also be supplied to meet the metabolic requirements for body maintenance. Feed requirements are greater for larger animals, obviously; however, the increase is not directly proportional to the increase in weight. Feed requirements for maintenance expressed as a percent of the live weight of mature dairy cattle decreases slightly for larger and larger animals. For example, cows weighing 800, 1000, 1200, 1400, and 1600 lb require approximately 1.48, 1.40, 1.34, 1.32, and 1.30 percent of their weights, respectively, in feed for maintenance. An 800-lb cow needs about 11.8 lb of feed where as the 1600-lb cow requires, not 23.6 lb (which would be twice the amount for the 800-lb cow), but only about 20.8 lb. The assumption must be made that the amounts of feed in any case will contain all of the required nutrients. The requirements for energy and protein expressed as a percentage of the live weight follow a similar pattern. The TDN requirement for 800- and 1600-lb cows is about 6.3 and 9.8 lb, respectively. The digestible protein requirements for cows of those weights are approximately 0.51 and 0.87 lb, respectively. Again the amount required for the 1600-lb cow is less than twice the amount needed by the 800-lb cow.

The need for nutrients in growing animals increases considerably over that required just to maintain an animal. Growing animals are involved in the production of additional tissues throughout the body, particularly muscle and bone. The protein requirements tend to increase at a rather rapid rate in growing animals. Whereas the comparison between a 1000-lb growing dairy heifer and a 1000-lb mature cow on a maintenance ration may not be completely valid it will give us some idea of the differences in requirements for some nutrients. The cow on a maintenance ration requires about 0.6 lb of protein, the growing animal needs about 1.0 lb. The increased requirement in the growing animal represents about a 67 percent difference. Mineral requirements, especially for calcium and phosphorus increase quite a bit also. Energy requirements increase from about 7.1 lb of TDN for the cow on a maintenance ration to approximately 10.2 lb of TDN for the growing animal. The requirements of other nutrients would be expected to increase accordingly. The requirements of finishing and fattening animals are similar to those of other growing animals. Particular nutrient needs will vary with the rates of growth and/or finishing.

Lactating animals have increased need for energy, protein, calcium, and most of the other nutrients. Milk contains protein, fat, sugar, and calcium in relatively large amounts when compared to the total solids content. The requirements for these nutrients in the diet of lactating animals will rise to compensate for that which is transferred to the milk. Note in Table 13-5 the TDN and digestible protein requirements (24.0 and 3.1 lb, respectively) of the 1200 lb cow producing 50 lb of milk. The requirements for these two nutrients for a 1200-lb cow on a maintenance ration is only 8.8 and

TABLE 13-5. ENERGY AND PROTEIN REQUIREMENTS
OF SELECTED FARM ANIMALS

Animal	Pounds of Feed (approximate)	Pounds of Total Protein	Pounds of Digestible Protein	Pounds of TDN	Megacalories of Digestible Energy (DE)
Beef cows, dry, 1100 lb	21.0	1.12	0.53	9.9	20.0
Beef steers, finishing, 550 lb, 2.0 lb/day gain	14.7	1.52	0.97	9.9	19.8
Dairy cows, 1200 lb, 50 lb of 3.5% milk	—	5.2	3.1	24.0	48.1
Pigs, finishing, 150 lb	6.8	0.9	0.8	5.3	10.4
Boars, 320 lb	5.7	0.77	0.63	4.2	8.3
Lactating mares, 1300 lb	24.0	3.1	1.9	15.0	30.0
Finishing lambs, 80 lb	3.5	0.35	0.2	2.2	4.1

Source: Adapted from National Research Council data.

0.7 lb, respectively. The requirement for calcium increases from 14 g per day for the cow on a maintenance ration to 74 g for the lactating cow—again, a considerable increase. The increases for these nutrients are not so dramatic for lactating animals of other species since they do not produce nearly so much milk. Because of the increased demand for nutrients for lactating animals, the structure of their diets should be adjusted accordingly during this period of time.

Very high producing dairy cows will sometimes develop milk fever following parturition if calcium supplies are inadequate to meet their high requirements for lactation. The sudden demand for calcium at calving results in a drain of the bone and blood supplies of that mineral. The deficiency in the blood affects the supply to the muscles. Since the muscles depend in part upon calcium for normal activity, the calcium deficiency prevents the cow from being able to stand up. The problem can be remedied with injections of calcium gluconate and ultimately by increasing calcium supplementation.

Growing and lactating animals also have increased need for vitamin A and the B-vitamins as well as the minerals. Producers working with the various species of domestic animals should obtain complete information concerning the nutrient requirements of their animals. Such information is available for any species of domestic and laboratory animal from the National Academy of Sciences–National Research Council. The material in Tables 13-1 and 13-5 was adapted from National Research Council (NRC) data.

The additional requirements of pregnant animals beyond maintenance needs are similar to the requirements of any other young growing animal. This is particularly true during the last one-third of the gestation period. Even though pregnant animals have an increased demand for protein and other nutrients, this demand is not as great as that of lactating females. In mares, the crude protein requirement for 1300-lb mares at very light work is about 1.1 lb per day. For pregnant mares of the same weight group the requirement is about 1.5 lb during the last third of pregnancy. However, the lactating mares require about 3.1 lb of total protein daily (see Table 13-4) during peak production. The requirements for energy, vitamin A, calcium, phosphorus, and the other minerals and vitamins also increase during pregnancy. The physiology associated with pregnancy is such that if pregnant females do not receive nutrients sufficient enough to supply the needs of the fetus, they can be withdrawn from the body of the dam. If the deficiencies are great enough, both the fetus and the dam will eventually suffer. Because the additional nutrient requirements for the production of viable sperm and ova by males and females are so small, they are usually not a major nutritional factor. This is true if the breeding animals are receiving adequate nutrients to supply their needs for body maintenance and any other work or production activities.

Animals involved in physical activity beyond the normal maintenance functions will require additional energy-producing nutrients. Working animals such as horses, breeding males, and hunting dogs would fit into this category. Work of this nature involves primarily additional muscular activity. This calls for additional energy. A 1200-lb horse at light work requires about 11.8 lb of TDN daily. The same horse at medium work will require about 13.8 lb or 2 lb more energy. The protein requirement, however, changes very little if any between these two levels of activity.

Ration Balancing

There is sometimes some confusion between the meanings of the terms "ration" and "diet." Diet refers to the usual food and drink for a person or animal. No mention is made in this particular definition as to the period of time covered by the diet, although some define it as the food and drink taken during one day. Ration is more specifically defined as the *daily* allowance of food for an animal. In this book, the two terms will be considered as synonymous. Neither term suggests that the food consumed contains all of the needed nutrients or that the amount is ample to satisfy the animal's needs. But adding the term "balanced" before either diet or ration does suggest that the amounts of feed and nutrients are allotted to meet the animal's requirements for the particular activity being undertaken. Our attention will now be turned to some of the procedures used to prepare balanced rations or balanced diets for domestic animals. Similar procedures can be used to prepare balanced diets for humans.

One procedure involves the supplying of nutrients in order of their magnitude of need. In most rations energy nutrients are required in the largest amount compared to other nutrients. Protein requirements are the second largest. From there, individual vitamins and minerals such as calcium, phosphorus, and vitamin A are considered. The best way to explain the method is to work through an example. Assume that we are to prepare a balanced ration for an 550-lb finishing steer. Note the TDN and digestible protein (DP) requirements in Table 13-5. With cattle and other herbivorous animals, an attempt should be made to supply as much of the energy (TDN) as possible from a roughage feed. A roughage would most likely provide the most economical source of nutrients. Suppose that orchardgrass hay is chosen as the roughage. Note that the total feed requirement for the steer is 14.7 lb. Calculating the amount of TDN that would be provided by 14.7 lb of orchardgrass hay we find $14.7 \times .53 = 7.89$ lb of TDN. Since the TDN requirement for the steer is 9.9 lb, there is a deficiency of 2.11 lb (9.9 − 7.89) of TDN. The next step is to choose a high energy feed to substitute for some of the hay to balance the TDN requirement. Suppose that corn is chosen as that feed. For every 1 lb of corn that is substituted for each 1 lb of hay, the TDN content of the ration will be increased by 0.25 lb (0.78 lb of TDN in corn – 0.53 lb TDN in hay = 0.25 lb difference). Since the ration is 2.11 lb deficient, a substitution of 2.11 ÷ 0.25 or 8.44 lb of hay must be replaced with that same amount of corn. The ration will now contain 8.44 lb of corn and 6.26 lb of hay (14.7 - 8.44 = 6.26). Checking the balance of TDN and digestible protein we find

	Feed (lb)	DP (lb)	TDN (lb)
Hay	6.26	0.376	3.32
Corn	8.44	0.549	6.58
Total feed	14.70	0.925	9.90

The TDN is balanced but the DP is 0.045 lb below the 0.97-lb requirement for the steer. The next step is to choose a high-protein supplement, say, soybean oil meal (SBOM), to supply the needed protein. For every 1 lb of SBOM substituted for each 1 lb of corn, there will be an increase of 0.325 lb of DP in the ration (0.39 lb DP in SBOM – 0.065 lb DP in corn 0.325 lb difference). It will be necessary to replace 0.14 lb of corn with an equal amount of SBOM to supply the 0.045 lb of needed protein (0.045 ÷ 0.325 = 0.14). Again checking the balance of TDN and DP we find

	Feed (lb)	DP (lb)	TDN (lb)
Hay	6.26	0.376	3.32
Corn	8.30	0.539	6.47
SBOM	0.14	0.055	0.10
Total feed	14.70	0.970	9.89

The ration is balanced to meet protein and energy requirements. The next step would be to check the content of vitamin A, calcium, and phosphorus. As it happens, this ration contains adequate amounts of these three nutrients. Many producers supply minerals to their livestock by providing free-choice, a mixture containing trace-mineralized salt and either steamed bone meal or dicalcium phosphate. Small amounts of these minerals can also be provided in a complete mixed ration. Because vitamin A can be stored in the liver for long periods of time, doses of this nutrient large enough to last for several months can be fed or injected into the animals.

Balancing a ration for swine and poultry would involve procedures similar to those explained for cattle except that roughage would not make up the primary part of the ration. The energy requirement would be met by supplying a high-energy feed. Any deficiency in the protein requirement would then be met by providing a protein supplement. Finally the vitamin and mineral requirements would be supplied with appropriate supplementation.

Another important consideration in the balancing of horse, swine, and poultry rations is the matter of protein quality. A protein supplement of animal origin or soybean oil meal must be chosen. The amino acid balance of the other plant proteins is unsatisfactory in meeting the amino acid requirements of nonruminant animals. The B-vitamins must also be supplied by the feed or supplemented for nonruminants. The microorganisms in the digestive systems of ruminants synthesize B-vitamins in adequate amounts to meet their requirements; this is not true in nonruminants.

STUDY QUESTIONS AND EXERCISES

1. Define each of the following terms:

air-dry	feces
balanced ration	feedstuff
basal feed	fodder
bulk	food
calorie	forage
calorimeter	gluten
coefficient of digestibility	grass
concentrate	graze
crude fiber	greenchop
crude protein	gross energy
diet	hay
diethylstilbestrol	heat increment
digestibility	implant
digestible energy	Kjeldahl
digestible protein	legume
energy	lignin
ether extract	meadow

metabolizable energy   silage
milk fever   stilbestrol
net energy   stover
NFE   straw
oil meal   supplement
pasture   tankage
protein supplement   TDN
proximate analysis   therm
ration   total digestible nutrients
roughage   urea

2. What feature distinguishes the roughages from other feedstuff materials?

3. What are some materials included among the plant fibers?

4. Name at least three major differences between grasses and legumes.

5. Name three factors that affect the nutrient content of a given forage crop.

6. Compare the water content of several kinds of forage.

7. Compare the ways in which different types of forage crops are harvested, stored, and fed.

8. What are some examples of feeds that are used primarily to provide energy for the diet?

9. What are some of the characteristics of high-energy feeds?

10. Give several examples of protein supplements of plant origin.

11. Name several protein supplements of animal origin.

12. Which parts of seeds usually contain the protein?

13. How is the water content of a feedstuff determined?

14. How is the fat content of a feed determined? What problems are associated with the procedure?

15. How is the fiber content of a feed determined?

16. How is the protein content of a feed determined?

17. Explain how a digestion trial is conducted.

18. When calculating the TDN content of a feed, why is the digestible fat fraction multiplied by 2.25?

19. How is the digestible energy (DE) value of a feed determined?

20. What difference is noted in determining the metabolizable energy value of a feed for ruminants compared to nonruminants?

21. What are some ways in which energy is lost from an animal's body?

22. Compare the relative protein and energy requirements for growing, working, lactating, and pregnant animals.

23. What happens in a pregnant animal if its nutrient intake is not sufficient to meet both its requirements and those of its fetus?

24. Why is it important to utilize roughages in balancing rations for cattle and sheep, while it is not an important consideration in balancing swine rations?

25. What factors must be considered in balancing rations for non-ruminants that need not be considered in the balancing of ruminant rations?

PROBLEMS

1. Assume that an air-dry sample of wheat weighs 74 g. After the sample was heated for several hours in an oven at 100°C, the weight changed to 65 g. Calculate the percent water and percent dry matter content of the sample.

2. Assume that the nitrogen content of a sample of soybean meal was determined to be 7.2 percent. What is the crude protein content of the feed?

3. Suppose that a proximate analysis of barley straw yielded these results: 10 percent water, 3.7 percent protein, 1.6 percent fat, 37.7 percent fiber, and 6 percent ash. What is the NFE content of this forage?

4. Show how the factor of 6.25, which is used to estimate the percent crude protein in a feed, was obtained.

5. If the crude protein content of a 100-g sample of sweet clover silage is 7.9 g and the protein content of the feces collected in a digestion trial involving the feeding of this silage were 1.9 g, what is the percent digestible protein content of the sample?

6. What is the coefficient of digestibility for the protein in the sample of sweet clover silage in problem five?

7. Assume that a digestion trial for a steer being fed 7 kg of alfalfa hay daily yielded the following results:

	Sample weight (g)	Water (g)	Ash (g)	Nitrogen (g)	Fiber (g)	Ether extract (g)
Hay	7000	634	560	172	2002	133
Feces	13,000	10,487	235	50	1101	93

    a. Calculate the percentages of crude protein, fiber, fat, and NFE.

    b. Calculate the percentages of digestible protein, fiber, fat, and NFE in the hay.

    c. Calculate the coefficients of digestibility for protein, fiber, fat, and NFE.

    d. Calculate the TDN for the hay.

8. Suppose that the caloric values of various heat losses in the above digestion trial were 15.4 Mcal for the feces, 1.4 Mcal for gases, 1.3 Mcal for urine, and 3.9 Mcal for the heat increment while the gross energy in the 7 kg of feed was 30.8 Mcal. Calculate the GE, DE, ME and NE in Kcal/kg.

9. Balance the digestible protein and TDN requirements in a ration for a 1300-lb lactating mare using barley, bromegrass hay, and alfalfa meal if needed; use the TDN content suggested for cattle.

10. Balance a ration for a 150-lb finishing pig for total protein and digestible energy (DE), using corn and tankage. Suggestion: Balance for protein first.

11. Prepare a balanced ration for an 80-lb finishing lamb using ground milo, red clover hay, and cottonseed meal if needed. Balance for total protein and TDN using the TDN content suggested for cattle.

12. Balance a ration for an 1100-lb dry cow using your choice of feeds. Balance for either TDN or DE, and either total or digestible protein.

SOLUTIONS TO PROBLEMS

1. 12 % water, 88% dry matter.
2. 45%.
3. 41%.
4. 100% protein ÷ 16% nitrogen = 6.25
5. 6%.
6. 76%.
7. a. 15.4%, 28.6%, 1.9%, and 37.1%.
   b. 10.9%, 12.9%, 0.57%, and 26.1%.
   c. 71%, 45%, 30%, and 70%.
   d. 51.1%.
8. 4400, 2200, 1814, and 1257.
9. 13.6 lb of barley, 9.75 lb of hay, and 0.65 lb of alfalfa meal.
10. 6.1 lb of corn and 0.7 lb of tankage.
11. 1.58 lb of hay and 1.92 lb of milo.
12. 21 lb of any hay from Table 13-1, or 50 lb of corn silage, or 55 lb of sorghum silage.

14

Meat and Wool

Animals produce a number of products that are useful to humans. The major use of animal products is for food. These food products include eggs, milk and other dairy products, poultry meat, and red meat. In addition to food products, animals produce many other important materials such as wool, leather, and drugs. Some animals are also used for work and pleasure. This chapter will deal primarily with the food products from meat animals and wool from sheep. Poultry and dairy products will be discussed in the next chapter.

LIVESTOCK MARKETING

The agricultural animal industry can be separated into several divisions such as production, marketing, and processing. Animal production is associated primarily with the activities that take place on farms and ranches and in feedlots. Livestock marketing as it relates to beef cattle, sheep, and hogs begins when the animals leave the farm and enter the marketing channel. It includes transportation, marketing organizations and services, information services,

314

and even some aspects of processing. Processing of meat animals includes the slaughter and preparation of the carcasses for sale to wholesale and retail outlets.

Importance of Livestock Marketing

Something on the order of 250 million animals are marketed each year in the United States. About one-half this number is cattle. Think of the vast numbers of people employed in the handling, transportation, and processing of these animals. The value of the marketing transactions involving these animals amounts to approximately $30-$40 billion each year. About three-fourths of this amount comes from cattle and calves.

An important facet of the whole marketing process involves the problem of distribution of the products. For example, only about one-half of the beef cattle in the United States are produced east of the Mississippi River, but almost three-fourths of the beef is consumed there. The western part of the country produces a surplus of beef that must be shipped east. The northeastern part of the country in particular is a beef deficit area. The production and consumption of sheep and lambs tend to follow a pattern similar to that for beef. Less than one-third of the lamb production in the country is in the east, yet over three-fourths of the consumption occurs there. Most of the hogs are produced in the central part of the country and must be shipped both east and west.

Methods of Marketing Livestock

Several methods of marketing livestock for meat purposes exist including terminal markets, local auctions, direct marketing, and selling on a grade and yield basis. These are the major ways in which livestock are marketed although many other methods are used, some of which are variations and modifications of those mentioned.

The local auction market is often referred to as a sales barn, community sale, or community auction. It is a trading center where livestock are bought and sold at public auction by the highest bidder. They are generally smaller than terminal markets and are usually owned by one person or firm. The seller pays a small charge for yardage, feed, and other services.

Terminal markets are usually rather large trading centers that include a stockyards company and several commission firms. This type of market is also referred to as a central market. The stockyards company could be thought of as the physical plant which includes the holding pens, building, loading chutes, and the office building. The stockyards company usually takes no active part in the actual buying, handling, and selling of the animals.

The buying, selling, and handling of livestock is done by the commission firms or companies. Producers or sellers consign their animals to a commission firm whose responsibility is to find a buyer and obtain the best possible price for the seller. The commission firms lease office and pen space from the stockyards company and collect a commission and service fee from the seller. The cost to the seller generally runs from one to five dollars per head depending upon the type and class of animals being sold. The cost to the seller is generally a little less than that of selling through a local auction market.

Two methods of selling are practiced at the terminal market: private treaty and auction. In the private treaty method, order buyers deal directly with representatives of the commission firms on a one-to-one basis. An order buyer is someone who buys for another person or company such as a feedlot or meat processing plant. The auction method is similar to that practiced at the local auction markets. Most of the terminal markets now operate their own auction rings where order buyers and other members of the public may bid (see Fig. 14-1).

The percentage of all livestock sold through terminal markets has been decreasing in favor of direct selling and local auctions. However, because most of the price quotations made by the marketing news services are based upon sales at the terminal markets, the impact of the terminal market upon livestock pricing is quite significant. Among the leading terminal market stockyards in the United States are those at Omaha, St. Paul, Oklahoma City, and Sioux City. Others include Joliet, Peoria, and National Stockyards, Illinois; St. Joseph, Kansas City, and Springfield, Missouri; and Amarillo and Ft. Worth, Texas. The leading sheep market is

Figure 14-1   Auction ring at a public livestock market. (Courtesy Duane Dailey, University of Missouri)

at San Angelo, Texas. A terminal market stockyards is shown in Figure 14-2.

The direct selling method of livestock marketing involves the sales of animals by the producer directly to the packer, local dealer, or other party. It does not involve a commission firm or selling agent of any kind. The buyer and seller negotiate to determine the price. Information about prices obtained for livestock sold by the direct method is difficult to obtain because of the private nature of the sale. This method of selling allows livestock to move in a more direct route from the producer to the packer or feedlot, and tends to reduce the overall cost of marketing. A majority of all livestock sold in the United States is handled by some variation of direct marketing.

Selling on the basis of carcass weight and grade is a way for farmers to be paid for producing better-quality slaughter animals. With this procedure, the live animals and the carcasses must be

Figure 14-2   A terminal market stockyard. (Courtesy Kansas City
Stock Yards Company)

identified as to the owner. The seller is paid based upon the actual
weight of the carcass and the final USDA carcass grade rather than
just on the live weight of the animals. By this method of selling,
producers can design their breeding and feeding programs to produce
the grade and weight of carcass that will bring them optimum
returns.

Market Classes of Livestock

In order to make the marketing of livestock more efficient, the
animals must be classified and graded. Market classes are based upon
distinctions involving species, use, sex, weight, and age. The quality
grade depends upon the degree of excellence of the animals for a
particular use.

TABLE 14-1.   MARKET CLASSES OF CATTLE
AND CALVES

	Cattle Use Classes		
	Slaughter	Feeder	Milkers and Springers
Sex classes	Steers Heifers Cows Bullocks Bulls	Steers Heifers Cows Bullocks Bulls	Cows
Age	Yearlings Two years and older	Yearlings Two years and older	All ages
	Calf Use Classes		
	Slaughter	Feeder	Vealers
Sex classes	Steers Bulls Heifers	Steers Bulls Heifers	No sex classes
Age	Three months to one year	Six months to one year	Under three months

Market classes of cattle and calves are shown in Table 14-1. Animals younger than one year are usually called calves while those older than one year are called cattle; however, the distinction is not clear-cut. Sometimes, very large animals that are younger than one year may be called cattle. Use classes of cattle include slaughter cattle, feeder cattle, and milkers and springers. Springer cattle are those which are about to have calves. Within the slaughter and feeder cattle classes, further distinction is made based upon sex and age. Steers and heifers are grouped according to yearlings and those two years and older. Bullocks are bulls under two years of age.

Use classification of calves include vealers which are calves of either sex under three months of age. Slaughter calves include steers, bulls, and heifers three months to one year of age. Feeder calves include those from weaning age (about six months) to one year of age, and also include bull, heifer, and steer groups.

The classification of sheep and lambs is similar to that for cattle and calves. Lambs are animals under one year of age while those over one year of age are called sheep. Use classes of sheep include slaughter sheep, feeder sheep, and breeder sheep. Sex classes of slaughter sheep include ewes, wethers, and rams. Feeder and breeder sheep classes generally do not include rams. Spring lambs are born in early spring and sold before July first. If they are sold after July first they lose the spring lamb classification. Hothouse lambs are born in fall or early winter and sold at 6 to 12 weeks of age weighing from 30 to 60 lb. See Table 14-2 for market classes of sheep and lambs.

TABLE 14-2.   MARKET CLASSES OF SHEEP AND LAMBS

	Sheep Use Classes		
	Slaughter	Feeder	Breeding
Sex Classes	Ewes Wethers	Ewes Wethers	Ewes
Age	Yearling Mature	Yearling Mature	Yearling Two-, three-, or four- year-olds and older

	Lamb Use Classes		
	Slaughter	Feeder	Shearer
Sex Classes	Ewes Wethers Rams	Ewes Wethers	Ewes Wethers
Age	Hothouse lambs Spring lambs Lambs	All ages	All ages

Hogs under 120 lb or less than four months of age are normally referred to as pigs at the markets. The term hogs is reserved for the older and larger animals. Hogs and pigs are further classified into feeder and slaughter classes (see Table 14-3). Three sex classes of slaughter hogs are barrows and gilts, sows, and stags and boars. Little distinction is made between barrows and gilts for either the slaughter or feeder classes of hogs or pigs.

TABLE 14-3.   MARKET CLASSES OF HOGS AND PIGS

	Hog Use Classes	
	Slaughter	Feeder
Sex Classes	Barrows and gilts Sows Stags Boars	Barrows and gilts
	Pig Use Classes	
	Slaughter	Feeder
Sex Classes	Barrows, gilts, and boars	Barrows and gilts

RED MEAT AND MEAT PRODUCTS

In a broad sense, meat refers to the edible flesh of domestic and wild mammals, birds, and fish. Red meat refers to that of cattle, sheep, and hogs. Meat from cattle is called beef; that from very young calves is called veal. Meat from hogs is called pork. The meat from sheep is called lamb or mutton depending upon the age of the animal when slaughtered.

Meat as Food for Humans

The major function of meat as a food for humans is as a source of high-quality protein. Meat is also a good source of iron, phosphorus, thiamin, riboflavin, and niacin. Depending upon the amount of fat it contains, meat can also supply moderate amounts of energy. Table 14-4 lists the nutrient composition of several kinds of meat.

Red meat contains about 15 to 20 percent protein, 15 to 20 percent fat, and about 60 to 65 percent water. On a dry matter

TABLE 14-4. APPROXIMATE CONTENT OF SELECTED NUTRIENTS OF SEVERAL MEATS AND NUTRIENT REQUIREMENTS OF HUMANS

Meat (3 oz serving without bone)	Energy (Kcal)	Protein (g)	Iron (mg)	Thiamin (mg)	Ribo-flavin (mg)	Niacin (mg)
Beef roast, little fat	255	19	3	0.06	0.16	3.9
Hamburger pattie, lean	185	20	3	0.08	0.20	5.1
Lamb, leg roast	265	17	2.6	0.12	0.22	4.5
Ham, smoked	340	17	2.5	0.46	0.18	3.5
Pork chop	295	13	2.1	0.6	0.17	3.6
Chicken, broiled	115	17	1.4	0.06	0.15	10.5
Tuna, canned	170	21	1.2	0.04	0.10	10.9
Bologna	258	9	1.5	0.14	0.18	2.3
Nutrient Requirements						
Adult men	2500–3200	70	10	1.3–1.6	1.8	18–21
Adult women	1800–2300	58	12	1.0–1.2	1.5	17

basis, meat is approximately one-half to three-fourths protein. Meat contains all of the amino acids that are considered essential in human diets in more than adequate amounts. The essential amino acids for humans include all those that are essential in swine diets (see Table 11-2) except arginine and histidine. Because meat contains the eight essential amino acids in amounts that exceed human requirements, it is referred to as a complete protein. Other examples of complete proteins with regard to human nutrition are those found in eggs, milk, fish, and soybeans. For a more complete discussion of protein quality and the essential amino acids, refer to Chapter 11.

The average consumption of red meat in the United States is about 180 lb per person (see Table 14-5). Almost two-thirds of this is beef and veal, and about one-third is pork. A very small part, less than two lb per person, is lamb. During the past 25 years, the general trend in beef consumption has been upward. The annual per capita consumption of pork has not varied much since World War II. The general trend in lamb consumption during this period has been downward.

TABLE 14-5.   PER CAPITA CONSUMPTION OF RED MEAT IN POUNDS DURING THE 1970s

Year	Beef	Veal	Lamb and Mutton	Pork	All Meats
1970	113.7	2.9	3.3	72.7	192.6
1971	113.0	2.7	3.1	79.0	197.8
1972	116.1	2.2	3.3	71.3	192.9
1973	109.6	1.8	2.7	63.9	178.0
1974	116.8	2.3	2.3	69.1	190.5
1975	120.1	4.2	2.0	56.1	182.4
1976	129.3	4.0	1.9	59.5	194.7
1977	125.9	3.9	1.7	61.5	193.0
1978	120.1	3.0	1.6	61.4	186.1
1979	107.6	2.0	1.5	70.2	181.3

Source: *Agricultural Statistics*, 1980. USDA.

Meat Processing and Grading

Before animals are slaughtered, they must be stunned or rendered insensible by the use of a gunshot, captive bolt stunner, or carbon dioxide gas. The animals are then hoisted by their hind legs and bled. Cattle and sheep are skinned. Hogs are usually scalded and only the hair removed; however, an increasing number of hogs are now being skinned. The head and internal organs are removed and, in the case of cattle and hogs, the carcass is cut lengthwise through the vertebral column and separated into halves.

The weight of the carcass compared to the live weight of the animal varies with species. This is referred to as the dressing percentage or yield. Typical figures for dressing percent for hogs, cattle, and sheep are 70, 60, and 50, respectively. For purposes of comparison, dressed poultry usually represents about 80 percent of the live weight. Carcasses are usually hung in a cooler for several days to age and to allow the meat to become more tender. This tenderizing process is brought about by the action of the natural body enzymes. An enzyme solution may be injected just prior to slaughter to speed up the aging process after slaughter.

Carcasses are normally cut into parts called wholesale cuts. Figures 14-3, 14-4, 14-5, and 14-6 show the wholesale cuts as well as the retail cuts of beef, veal, pork, and lamb carcasses. Retail cuts are those that are normally found in grocery stores and meat markets. The highest priced cuts come from the loins and hind legs.

There is considerable similarity in many of the retail cuts of beef, veal, pork, and lamb as illustrated in Figure 14-7. For example, a rib steak from a beef carcass is very similar to a rib chop from a lamb or pork carcass. A leg chop from a lamb carcass is not too different from a fresh or smoked ham slice from a pork carcass or a round steak from a beef carcass. There is much similarity among the sirloin cuts of the several types of carcasses.

Some of the characteristics that have a large effect upon the palatability of meat are tenderness, flavor, and juiciness. The age of an animal at slaughter affects both tenderness and flavor, but in opposite ways. The meat from younger animals tends to be more tender while that from older animals tends to have a more desirable flavor. The goal of processing is to get the optimum flavor and tenderness. This means that the best carcass is not the one from a very young animal; although it would be the most tender it would probably lack flavor. Neither would the best carcass be obtained from the older animal which might have the maximum flavor but lack tenderness. The idea, then, is to select the medium-aged animals that will produce the optimum of flavor and tenderness.

Juiciness seems to be related to the amount of intramuscular fat present in the carcass; this is called marbling. Marbling apparently has some influence upon the degree of tenderness, particularly in older animals. Older carcasses seem to be juicier than younger carcasses. This factor could well be related to degree of marbling since younger animals usually do not show the degree of marbling of older animals. The amount of time on feed is a big factor in determining the degree of marbling.

Slaughter animals are graded based largely upon the degree and location of fat. Beef carcasses are assigned quality grades based upon the amount of marbling found in the loin muscle. Carcasses with the greatest amount of marbling are graded U.S. Prime. The next lower grade is U.S. Choice, then Good, Standard, Commercial, Utility, and Cutter and Canner. In addition, beef carcasses may be yield graded. Yield grades are based upon the percent of lean retail cuts that can be obtained from a carcass. The fatter the carcass, the lower the yield. Yield is similar to dressing percentage except that yield grades are expressed as numbers from 1 to 5; yield grade 1 is the leanest. Figure 14-8 includes several cuts of beef showing different yield and quality grades.

Lamb carcasses are graded in a manner similar to beef carcasses except that the quality grades are based more on the amount of fat between the ribs and over the flanks than on marbling. Quality grades for lamb are Prime, Choice, Good, Utility, and Cull. Lamb carcasses may also be graded according to yield. There are five yield grades as with beef carcasses.

A single grading system for slaughter hogs includes five grades based on both yield and quality. Two levels of quality are recognized based upon the firmness, marbling, and color of the loin. Those levels are either acceptable or unacceptable. If a carcass is judged as unacceptable with regard to quality it is graded U.S. Utility; this is the bottom grade. Acceptable carcass are then graded as either U.S. No. 1, 2, 3, or 4 depending upon the thickness of backfat and carcass length within certain weight groups.

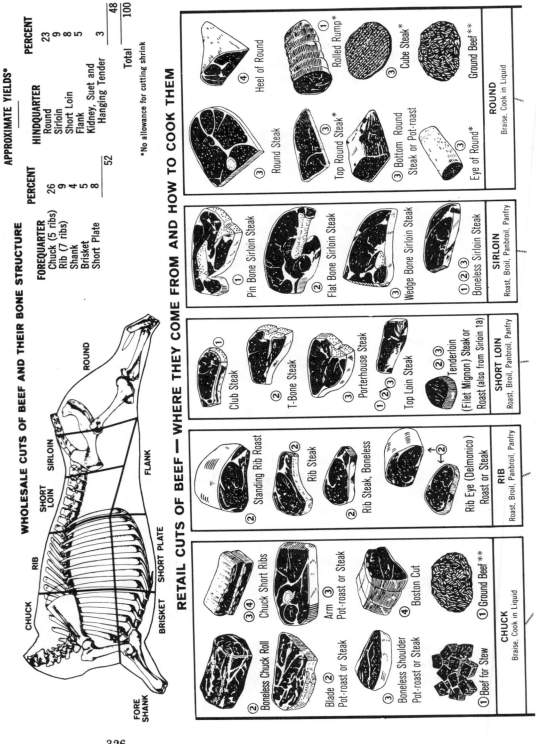

WHOLESALE CUTS OF BEEF AND THEIR BONE STRUCTURE

CHUCK · RIB · SHORT LOIN · SIRLOIN · ROUND · FLANK · SHORT PLATE · BRISKET · FORE SHANK

APPROXIMATE YIELDS*

FOREQUARTER	PERCENT
Chuck (5 ribs)	26
Rib (7 ribs)	9
Shank	4
Brisket	5
Short Plate	8
	52

HINDQUARTER	PERCENT
Round	23
Sirloin	9
Short Loin	8
Flank	5
Kidney, Suet and Hanging Tender	3
	48
Total	100

*No allowance for cutting shrink

RETAIL CUTS OF BEEF — WHERE THEY COME FROM AND HOW TO COOK THEM

CHUCK
② Boneless Chuck Roll
Blade ② Pot-roast or Steak
Boneless Shoulder Pot-roast or Steak
③④ Chuck Short Ribs
Arm ③ Pot-roast or Steak
④ Boston Cut
① Ground Beef**
① Beef for Stew
CHUCK Braise. Cook in Liquid

RIB
② Standing Rib Roast
② Rib Steak
② Rib Steak, Boneless
Rib Eye (Delmonico) Roast or Steak ← ②
RIB Roast, Broil, Panbroil, Panfry

SHORT LOIN
① Club Steak
② T-Bone Steak
③ Porterhouse Steak
① ② ③ Top Loin Steak
② ③ Tenderloin (Filet Mignon) Steak or Roast (also from Sirloin 1a)
SHORT LOIN Roast, Broil, Panbroil, Panfry

SIRLOIN
① Pin Bone Sirloin Steak
② Flat Bone Sirloin Steak
③ Wedge Bone Sirloin Steak
① ② ③ Boneless Sirloin Steak
SIRLOIN Roast, Broil, Panbroil, Panfry

ROUND
④ Heel of Round
Round Steak
② Rolled Rump*
③ Top Round Steak*
③ Cube Steak*
③ Bottom Round Steak or Pot-roast
③ Eye of Round*
Ground Beef**
ROUND Braise. Cook in Liquid

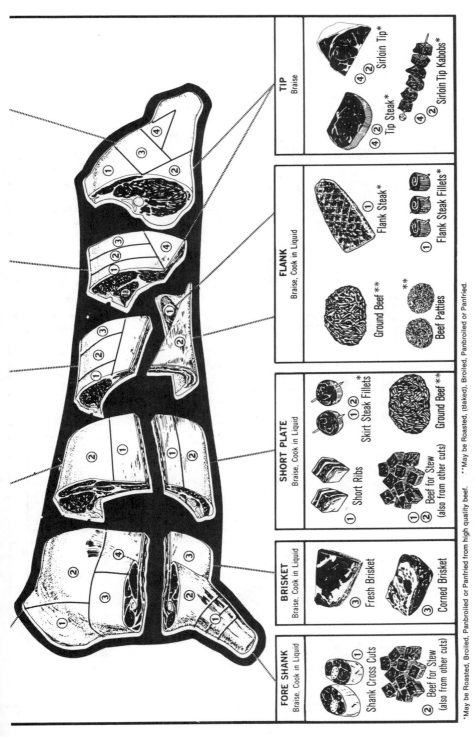

Figure 14-3 A beef chart. (Courtesy National Live Stock & Meat Board, Chicago, Illinois)

*May be Roasted, Broiled, Panbroiled or Panfried from high quality beef. **May be Roasted, (Baked), Broiled, Panbroiled or Panfried.

327

WHOLESALE CUTS OF VEAL AND THEIR BONE STRUCTURE

APPROXIMATE YIELDS*

NAME OF CUT	PERCENT
Shoulder	28.0
Rack, trimmed	7.3
Loin, trimmed	7.7
Leg (Sirloin on)	34.0
Shank (Trotter on)	3.8
Breast and Flanks	13.4
Kidneys and Suet	5.8
Total	**100**

*No allowance for cutting shrink

SHOULDER · HOTEL RACK · LOIN TRIMMED · LEG · FLANK · BREAST · FORE SHANK

RETAIL CUTS OF VEAL — WHERE THEY COME FROM AND HOW TO COOK THEM

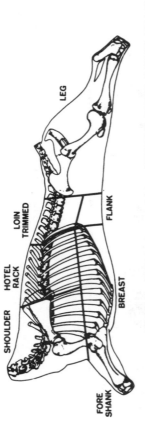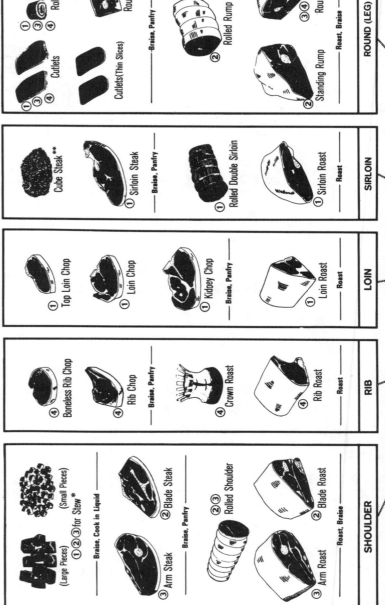

SHOULDER

(Large Pieces) (Small Pieces) ①②③ for Stew*
— Braise, Cook in Liquid —
③ Arm Steak
② Blade Steak
②③ Rolled Shoulder
— Braise, Panfry —
③ Arm Roast
② Blade Roast
— Roast, Braise —

RIB

④ Boneless Rib Chop
④ Rib Chop
— Braise, Panfry —
④ Crown Roast
④ Rib Roast
— Roast —

LOIN

① Top Loin Chop
① Loin Chop
① Kidney Chop
— Braise, Panfry —
① Loin Roast
— Roast —

SIRLOIN

Cube Steak**
① Sirloin Steak
— Braise, Panfry —
① Rolled Double Sirloin
① Sirloin Roast
— Roast —

ROUND (LEG)

Rolled Cutlets
①③④ Round Steak
④ Cutlets
Cutlets (Thin Slices)
— Braise, Panfry —
② Rolled Rump
② Standing Rump
③④ Round Roast
— Roast, Braise —

328

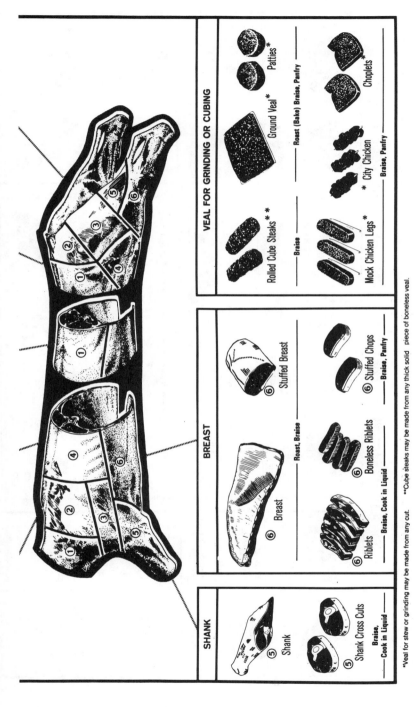

SHANK

⑤ Shank

Braise,
Cook in Liquid

Shank Cross Cuts

⑤

Braise,
Cook in Liquid

BREAST

⑥ Breast

Roast, Braise

⑥ Stuffed Breast

⑥ Boneless Riblets

⑥ Stuffed Chops

Braise, Panfry

Riblets ⑥

Braise, Cook in Liquid

VEAL FOR GRINDING OR CUBING

Rolled Cube Steaks **

Braise

Mock Chicken Legs *

Ground Veal *

* City Chicken

Braise, Panfry

Patties *

Choplets *

Roast (Bake) Braise, Panfry

*Veal for stew or grinding may be made from any cut.   **Cube steaks may be made from any thick solid   piece of boneless veal.

Figure 14-4 A veal chart. (Courtesy National Live Stock & Meat Board, Chicago, Illinois)

APPROXIMATE YIELDS*

NAME OF CUT	PERCENT
Fresh Hams, Skinned	21.0
Loins, Blade on	18.0
Boston Shoulder	6.6
Picnics, Shoulder	8.8
Bacon, Square Cut	17.3
Spareribs	3.8
Jowl, Trimmed	3.0
Feet, Tail, Neckbones	6.0
Fat Back, Clear Plate and all Fat Trimmings	11.2
Sausage Trimmings	4.3
Total	100

*Packer Dressed Hog, Head off, Leaf out
No allowance for cutting shrink

WHOLESALE CUTS OF PORK AND THEIR BONE STRUCTURE

HAM (LEG)

HIND FOOT

LOIN

BACON (BELLY)

FAT BACK

CLEAR PLATE

SPARERIBS

BOSTON BUTT

PICNIC

TRIMMED JOWL

FORE FOOT

RETAIL CUTS OF PORK — WHERE THEY COME FROM AND HOW TO COOK THEM

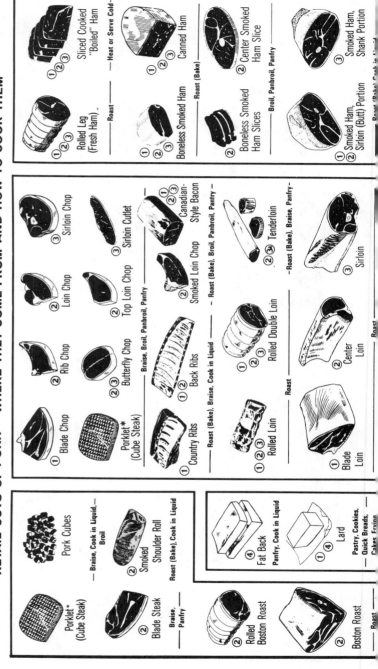

① ② ③ Rolled Leg (Fresh Ham) — Roast —

① ② ③ Sliced Cooked "Boiled" Ham — Heat or Serve Cold —

① ② ③ Boneless Smoked Ham — Roast (Bake) —

① ② ③ Canned Ham — Roast (Bake) —

② Boneless Smoked Ham Slices

② Center Smoked Ham Slice — Broil, Panbroil, Panfry —

① Smoked Ham, Sirloin (Butt) Portion

③ Smoked Ham, Shank Portion

③ Sirloin Chop

② Loin Chop

② ③ Butterfly Chop

① Blade Chop

Porklet* (Cube Steak)

① Country Ribs

③ Sirloin Cutlet

② ③ Top Loin Chop

— Braise, Broil, Panbroil, Panfry —

① ② Back Ribs

② Smoked Loin Chop

— Roast (Bake), Broil, Panbroil, Panfry —

① ② ③ Rolled Loin

— Roast (Bake), Braise, Cook in Liquid —

② ③ ① Canadian-Style Bacon

② ③ ④ Tenderloin

① ② ③ Rolled Double Loin

② Center Loin

① Blade Loin

— Roast —

② ③ Sirloin

— Roast —

Pork Cubes
— Braise, Cook in Liquid, Broil

② Smoked Shoulder Roll
Roast (Bake), Cook in Liquid

Porklet* (Cube Steak)
— Braise, Panfry —

② Blade Steak

② Rolled Boston Roast

② Boston Roast
— Roast —

④ Fat Back
Panfry, Cook in Liquid

① ④ Lard
Pastry, Cookies, Quick Breads, Cakes Frying

330

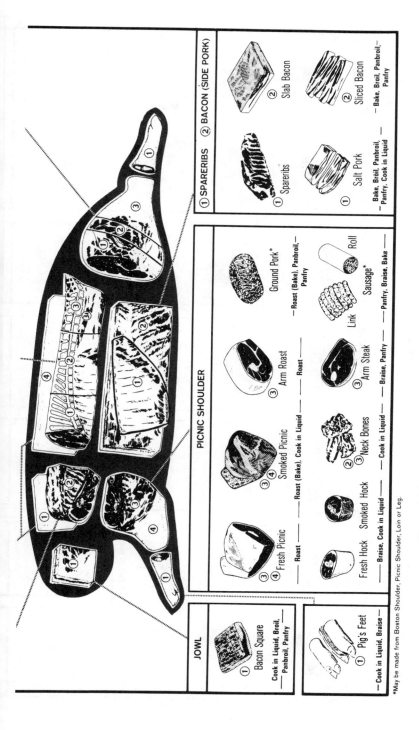

Figure 14-5 A pork chart. (Courtesy National Live Stock & Meat Board, Chicago, Illinois)

331

WHOLESALE CUTS OF LAMB AND THEIR BONE STRUCTURE

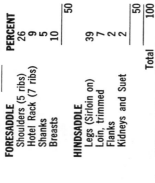

HOTEL RACK

LOIN TRIMMED

LEG

FLANK

BREAST

SHOULDER

FORE SHANK

*No allowance for cutting shrink

RETAIL CUTS OF LAMB — WHERE THEY COME FROM AND HOW TO COOK THEM

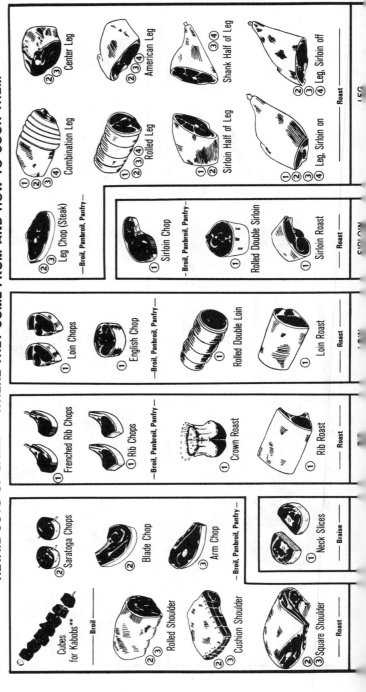

Center Leg ②③

American Leg ②③④

Shank Half of Leg ③④

Leg, Sirloin off ②③④

Combination Leg ①②③④

Rolled Leg ①②③④

Sirloin Half of Leg ①②

Leg, Sirloin on ①②③④

Roast — LEG

Leg Chop (Steak) ②③
—Broil, Panbroil, Panfry—

Sirloin Chop ①
— Broil, Panbroil, Panfry—

Rolled Double Sirloin ①

Sirloin Roast ①
Roast — SIRLOIN

Loin Chops ②
English Chop ①
—Broil, Panbroil, Panfry—

Rolled Double Loin ①

Loin Roast ①
Roast — LOIN

Frenched Rib Chops ①

Rib Chops ①
—Broil, Panbroil, Panfry—

Crown Roast ①

Rib Roast ①
Roast — RIB

Saratoga Chops ②

Blade Chop ②

Arm Chop ③
— Broil, Panbroil, Panfry —

Cubes for Kabobs** ①
Broil

Rolled Shoulder ②③

Cushion Shoulder ②③

Square Shoulder ②③
Roast — SHOULDER

Neck Slices ①
Braise

332

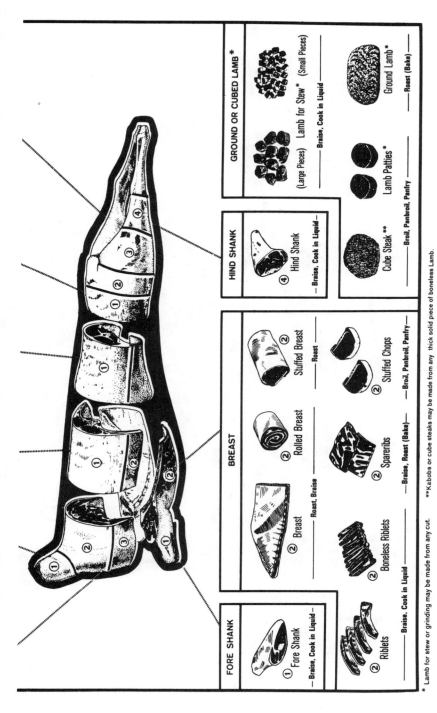

Figure 14-6 A lamb chart. (Courtesy National Live Stock & Meat Board, Chicago, Illinois)

GROUND OR CUBED LAMB *

(Large Pieces) Lamb for Stew * (Small Pieces)

— Braise, Cook in Liquid —

Ground Lamb *

— Roast (Bake) —

Lamb Patties *

— Broil, Panbroil, Panfry —

Cube Steak **

HIND SHANK

④ Hind Shank

— Braise, Cook in Liquid —

BREAST

② Stuffed Breast

— Roast —

② Rolled Breast

— Braise, Roast (Bake) —

② Breast

— Roast, Braise —

② Stuffed Chops

— Broil, Panbroil, Panfry —

② Spareribs

② Boneless Riblets

— Braise, Cook in Liquid —

FORE SHANK

① Fore Shank

— Braise, Cook in Liquid —

② Riblets

— Braise, Cook in Liquid —

* Lamb for stew or grinding may be made from any cut.   ** Kabobs or cube steaks may be made from any thick solid piece of boneless Lamb.

333

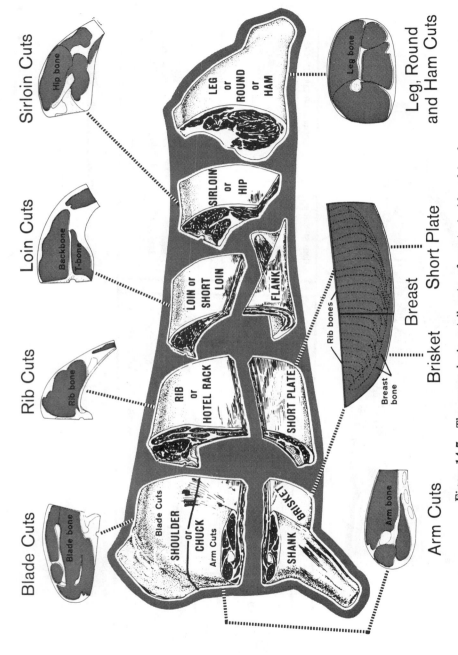

Sirloin Cuts

Hip bone

Loin Cuts

Backbone

T-bone

Rib Cuts

Rib bone

Blade Cuts

Blade bone

LEG
or
ROUND
or
HAM

SIRLOIN
or
HIP

LOIN or
SHORT
LOIN

FLANK

RIB
or
HOTEL RACK

SHORT PLATE

Blade Cuts

SHOULDER
or
CHUCK

Arm Cuts

BRISKET

SHANK

Leg bone

Leg, Round
and Ham Cuts

Rib bones

Breast
bone

Breast

Short Plate

Brisket

Arm bone

Arm Cuts

Figure 14-7 The seven basic retail cuts of meat. A side of beef appears in the chart as an example. It could just as well have been veal, pork, or lamb for comparison purposes. (Courtesy National Live Stock & Meat Board, Chicago, Illinois)

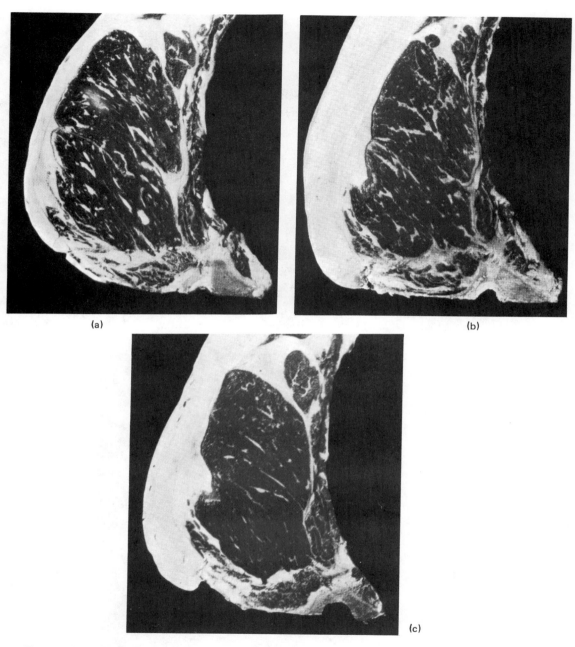

(a)

(b)

(c)

Figure 14-8 Beef ribeyes showing several quality and yield grades. Observe the decreasing amount of marbling in the muscle as you proceed from Prime to Choice to Good to Standard. Beginning with part (b) note the decreasing thickness of fat on the ribeye as you proceed from Yield Grade 5 through Yield Grade 1. (a) U.S. Prime, Yield Grade 2. (b) U.S. Choice, Yield Grade 5. (c) U.S. Choice, Yield Grade 4. (Courtesy National Live Stock & Meat Board, Chicago, Illinois)

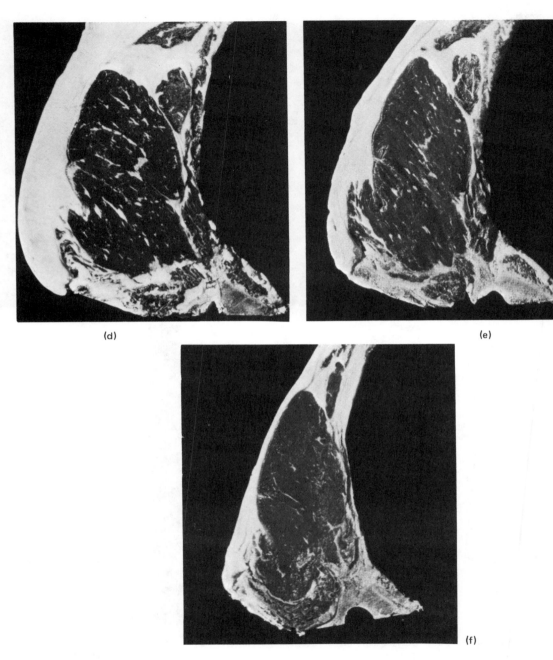

(d)

(e)

(f)

Figure 14-8   Beef ribeyes *continued*. (d) U.S. Choice, Yield Grade 3. (e) U.S. Good, Yield Grade 2. (f) U.S. Standard, Yield Grade 1. (Courtesy National Live Stock & Meat Board, Chicago, Illinois)

A number of so-called variety meats are produced from meat animals. Examples of variety meats include liver, kidneys, tongue, and brains. Other edible by-products are produced or manufactured from certain parts of the carcass. Animal fats such as lard and tallow are used to make a variety of foods such as oleomargarine, shortenings, candies, and spreads. Certain inedible products are also manufactured from fat such as lubricating oils, plastics, soap, explosives, candles, and a multitude of other products.

A number of by-products are used in the feed manufacturing industry. Some examples of meat by-product feeds are blood meal, meat and bone scraps, tankage, and liver meal. Fats are also often used in the manufacture of certain feeds. Various hormones such as thyroid extracts, pituitary extracts, estrogen, insulin, and adrenalin are among the long list of animal by-products. The list goes on to include such things as cholesterol, digestive enzymes, heparin (an anticoagulant), ingredients for cosmetics, and many kinds of leather goods.

WOOL

The natural hair of the sheep is called wool. It is similar to human hair except that it is smaller in diameter and is crimped. The crimp prevents the individual fibers from lying close together in the cloth. Because the fibers do not lie close together, about two-thirds to three-fourths of the volume of woolen fabric may be air. This feature gives wool a tremendous insulation value, not only as protection from the cold, but from the heat of the sun as well.

Wool is elastic. An individual fiber can be stretched up to 30 percent and then regain its original length. It can be crumpled and bent and return to its original shape. This property allows woolen garments that have become wrinkled to regain their shape if allowed to hang for a few hours. A wool fiber is stronger than a steel fiber of the same diameter.

Another important characteristic of wool is that it has a tremendous capacity to absorb water, up to 18 percent of its weight, without feeling damp. It can absorb moisture up to 50 percent of its weight without becoming saturated. This allows woolen garments to absorb perspiration and other dampness and keep it from clinging to the body of the wearer. Even though it is an organic substance, wool is almost nonflammable. It will burn only as long as it is exposed to a direct flame from another source.

Wool contains a protein called keratin which is the major constituent of hair, feathers, hoofs, horns, and finger and toenails. The fibers of wool vary in length from less than 1 in. to more than 6 in. The diameter varies from 15 to 50 μ (1/1700 to 1/500 in.). Fiber diameter is one of the factors used in the grading of wool. The smaller the diameter, the finer the grade, and the greater the number of yards of yarn that can be obtained from a pound of wool. The finest wool is obtained from the Merino breed. The Rambouillet breed produces the next finest wool. Most of the meat breeds produce a wool of medium fineness.

The USDA grading system refers to the number of hanks of yarn that can theoretically be woven from one pound of scoured wool. A hank of wool contains 560 yd of yarn. The grade numbers vary from about 36 to 80. The larger numbers represent the finer wool (see Fig. 14-9).

After the wool is removed from the animal it is sometimes called fleece. The weight of the fleece varies from about 5 to 11 lb with an average of approximately 8 lb per sheep. The raw product is called grease wool. Grease refers to all of the impurities in the wool including yolk, suint, and secretions of the sebaceous glands. Yolk is the yellow coloring in wool. Suint is produced by certain body secretions and carries the characteristic odor of the sheep. A common substance used in cosmetics called lanolin is obtained from wool grease.

The grease is removed from the wool by a process called scouring. This involves treatment with soap and other materials. The scoured fleece usually weighs less than one-half that of the original raw fleece. The total amount of shrinkage varies from 50 to 60 percent.

Two kinds of yarn are made from wool: worsted and woolen yarn. Woolen yarn is made from the coarser and shorter fibers, usually shorter than 2 in. The fibers lie in all directions and produce a thicker, fuzzier fabric such as that found in tweeds. Worsted fabrics are made from the longest, finest, and highest-priced wool, such as that produced from Merinos and Rambouillets. The long fibers lie parallel to one another producing a lighter fabric that has a smoother finish such as that in gabardine.

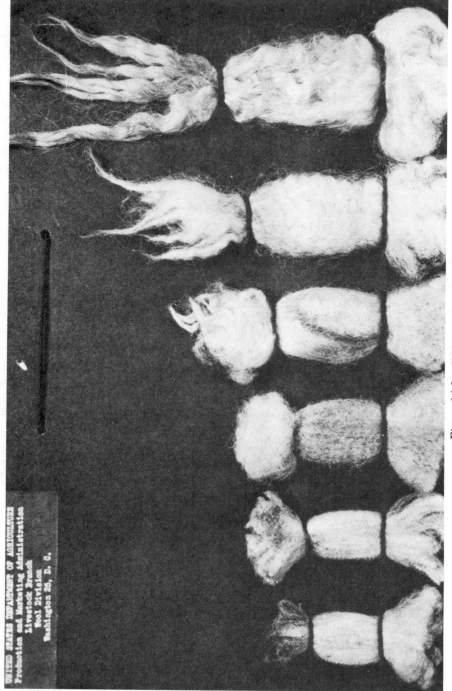

Figure 14-9 USDA grades of wool.

STUDY QUESTIONS AND EXERCISES

1. Define each of the following terms:

auction	mutton
carcass	order buyer
crimp	red meat
dressing percent	scoured wool
fleece	springer
flesh	stockyards
gabardine	suint
grease wool	tallow
hank	terminal market
keratin	veal
lanolin	vealer
lard	worsted wool
marbling	yield
meat	

2. What kinds of activities does livestock marketing include?

3. About how many animals are marketed each year and what is the approximate value of the transactions?

4. Which regions of the United States are surplus producers of beef, pork, and lamb?

5. Name at least four ways in which livestock can be marketed.

6. Compare a local auction market to a terminal or central market.

7. What is a commission firm and what is its function?

8. How are livestock sold at central markets?

9. Which method of marketing has the greatest impact upon pricing and why?

10. What are some of the advantages and disadvantages of selling livestock based on grade and yield?

11. What are the use classifications of cattle, sheep, and hogs?

12. Of which nutrients is red meat an excellent source?

13. Name the eight amino acids considered essential in human diets.

14. Name the two amino acids that are essential in the diets of swine but not essential in human diets.

15. Compare the annual per capita consumption figures for beef, pork, and lamb, and explain how they have changed since the 1950s.

16. Describe the slaughtering process for beef cattle and swine.

17. Compare the average dressing percent figures for sheep, swine, poultry, and beef cattle.

18. Which wholesale cuts are the most valuable in the three species of meat animals?

19. On a sketch of a beef carcass, locate and identify the various wholesale cuts. Do the same for a pork and lamb carcass.

20. Name the common retail cuts of meat that are obtained from the various wholesale cuts of beef, pork, and lamb carcasses.

21. What is the primary basis for grading slaughter beef carcasses?

22. Name the slaughter grades of beef, lamb, and pork carcasses.

23. Give some examples of variety meats.

24. Name several by-products obtained from the meat packing industry.

25. Explain why wool is a good insulator.

26. To what degree does wool absorb water and how is this an advantage?

27. How is the diameter of wool fiber related to quality?

28. Compare the characteristics of tweed and gabardine cloth.

29. About how much wool does a sheep produce?

15

Dairy and Poultry Products

Milk, eggs, and poultry meat play a major role in providing humans with many of the essential nutrients. Along with red meat and fish, they constitute the best source of high-quality protein for human consumption. The primary purpose of this chapter is to discuss the nutritional importance of dairy and poultry products. Because it is important to understand how animals produce these products, the biological processes of milk production and egg laying are also presented.

LACTATION

For a number of weeks after birth, about the only source of food for the young mammal is the milk from the dam. Because milk must contain the proper supply of nutrients to sustain the young animal during this period, it is an excellent source of many essential nutrients. Milk has developed a reputation as being one of nature's "most perfect foods". Most female mammals produce only enough milk to feed their young for a short time after birth. The dairy cow in particular is an exception in that, through years of selection, it has been developed into an extremely high producer of milk for human consumption.

The number of mammary glands present on an individual of any species is related to the number of offspring that are normally produced at one time. Mares, ewes, and women have 2 glands; cattle have 4. The number of glands present on animals that normally produce litters varies from 8 to 14.

Milk is produced in the mammary gland by millions of small saclike structures called alveoli. An alveolus is a sac made up of a single layer of cells. The milk is secreted inside the alveolus. All the components of milk are removed from the blood in the secretory process. In order to produce one unit of milk, 400 or more units of blood must pass through the udder. Each alveolus is surrounded by a network of blood capillaries and another network of smooth muscle fibers called myoepithelium. The muscle fibers function in forcing the milk from the alveolus.

Numbers of alveoli are connected to very small ducts to form a cluster of alveoli that are similar in appearance to a bunch of grapes. Several "bunches," or clusters, of alveoli are joined to another common duct to form a lobule; several lobules form a lobe of the mammary gland. Ducts from a number of lobes join together to form larger ducts. In a cow, several of the largest milk ducts lead to the gland or udder cistern (see Fig. 15-1). The gland cistern leads to the teat cistern. The milk is held inside these cavities until it is removed by the suckling of the calf or by the milking process.

In the udder of a cow, the lobule-alveolar network is supported by special connective tissue. The udder itself is suspended by sheet-like connective tissue called suspensory ligaments. For example, the medial suspensory ligament is situated such that it separates the right side of the udder from the left. The udder is designed in such a way that the four glands in the cow are functionally separate one from another. Mammary glands are very well supplied with blood vessels, nerves, and lymphatic vessels.

During each estrous cycle following puberty, estrogen stimulates some development of the duct system of the mammary glands. During pregnancy, progesterone stimulates further development of the duct system and especially development of the alveoli. As parturition approaches, the hormone prolactin from the anterior pituitary gland and cortisol from the adrenal cortex stimulate the secretion of milk itself. The stimulus of milking or nursing releases

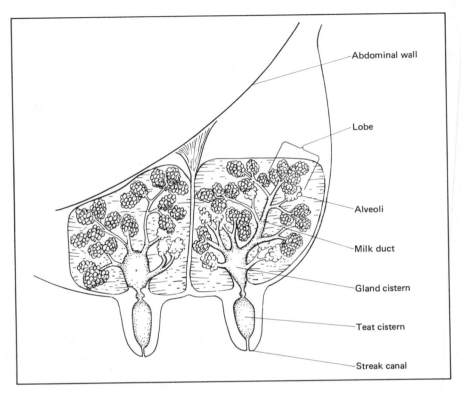

Figure 15-1   Diagram of a cow's udder.

oxytocin from the posterior lobe of the pituitary gland which results in the contraction of the myoepithelium of the alveoli thus forcing the milk down the milk ducts and into the gland cisterns. Because of this function, oxytocin is often called the milk let-down hormone. This is the same hormone that helps bring about the expulsion of the fetus during parturition.

MILK AND DAIRY PRODUCTS

Milk is a unique product of mammals. However, unless the term is qualified, it usually refers to the milk of the dairy cow. Well over 90 percent of the world supply of milk is produced from dairy cows.

A small amount of milk from other species is used for human food in the United States and other parts of the world. Goats, water buffalo, camels, sheep, reindeer, and llamas are among those animals that contribute milk for human consumption.

The annual consumption of milk in the United States is about 560 lb per person. Less than one-half of that is consumed as fluid milk, however. About 300 lb of it is converted to various dairy products such as butter, cheese, and ice cream. Trends in the consumption of dairy and poultry products since 1970 are shown in Table 15-1. The per capita consumption of all milk products has increased during the past few years after having decreased over a period of approximately 20 years. While it is not shown in the table, it is interesting to note that the consumption of those dairy products having the highest fat content has decreased most while some of those with lower fat content have actually increased in consumption.

TABLE 15-1. PER CAPITA CONSUMPTION OF DAIRY AND POULTRY PRODUCTS DURING THE 1970s

Year	Total Milk and Dairy Products (lb)	Dairy Products in Fluid Milk Equivalents (lb)	Eggs	Poultry Meat (lb)			
				Chicken	Turkey	Other	Total
1970	583	297	309	38.1	7.8	0.2	46.1
1971	581	302	312	38.2	8.0	0.3	46.5
1972	582	302	304	40.1	8.7	0.5	49.4
1973	559	282	291	39.9	8.6	0.3	48.8
1974	552	292	286	40.3	8.8	0.2	49.3
1975	546	286	280	39.9	8.1	0.3	48.3
1976	566	304	274	44.5	9.2	0.2	53.9
1977	572	311	272	45.4	8.8	0.3	54.5
1978	564	307	278	48.0	9.2	0.3	57.5
1979	567	309	284	52.5	10.0	0.4	62.9

Source: *Agricultural Statistics*, appropriate years. USDA.

Milk Production

About three-fourths of all milk produced on farms is sold as Grade A milk. This refers to milk that has been produced under specific sanitary conditions that make it acceptable for sale and use as fluid milk. Milk designated as Grade B or Grade C is produced under less stringent conditions that do not allow it to qualify as Grade A. Grade B and Grade C milk is converted to manufactured products such as butter, cheese, and nonfat dry milk. Grading standards are developed by the U.S. Department of Agriculture. Class I and Class II are use categories for milk. Class I milk is that which is sold in bottles or cartons as fluid milk. Only Grade A milk can be Class I. Class II milk is any milk that is used as manufacturing milk whether it be Grade A or Grade B. Surplus Grade A milk can be put to Class II use, but it is still Grade A milk.

Most of the milk sold commercially in the United States is pasteurized. This is a process whereby the milk is exposed to temperatures high enough to kill all pathogenic organisms. The process used today is to heat milk to a temperature of 161°F (72°C) for 15 seconds. Some years ago, the pasteurization process involved a temperature of 145°F and a time of 30 minutes.

Another process to which milk is usually exposed is that of homogenization. This is a process whereby the fat globules are broken into smaller droplets or globules so they will remain dispersed in the milk and not rise to the top. It is performed by forcing milk through very small openings at high pressure or exposing it to high frequency sonic vibrations. Homogenized milk is an emulsion of the butter fat with the nonfat portion of the milk.

Nutritional Value of Milk

Milk is an excellent source of a number of nutrients important in human nutrition. These include protein, calcium, and riboflavin in particular. One quart of milk will supply one-half the daily protein needs of an adult and essentially all the calcium and riboflavin requirements. Milk is also a good source of phosphorus, other B-vitamins, vitamin A, and energy. Milk is not a very good source of iron and vitamin C.

The key factor in the importance of milk in human nutrition is not so much the quantity of protein or vitamins or minerals it contains, but the quality of the protein. The protein of milk contains all of the amino acids needed in human diets and in more than ample quantities to meet those needs. The quality of milk protein is second only to that of eggs.

The percentage of protein in milk varies from around 5.0 in the ewe and sow to about 1.5 in women. The milk of cows and goats contains about 3.5 percent protein while that of the mare contains about 2.5 percent protein.

Milk is one of the few foods that contains ample amounts of the three compounds: protein, fat, and carbohydrate. The amounts of these substances very with species, breeds, and individuals. Among farm animals and humans, the amount of fat in the milk varies from about 1.5 percent for the mare to about 8.0 percent for the ewe. Sow's milk contains about 5.0 percent fat. The amount of fat in the milk of cows, goats, and women is about 4.0 percent.

In dairy cattle the fat content is quite variable among breeds as well as individuals. It is dependent upon such factors as age, stage of lactation, nutrition, and breeding. Fat percentage in cattle varies from as low as 3.0 to over 5.0. Generally, the Jersey and Guernsey breeds produce the highest percentage of fat while the Holstein produces the lowest. This does not mean that Holsteins produce the least total fat, however. Because they produce so much more total milk than the Jersey and Guernsey, they also usually produce as much or more total butterfat. In fact, the record for the most butterfat produced in one year is held by a Holstein cow.

The sugar of milk is called lactose. Lactose is a compound sugar, or disaccharide, which has been found almost nowhere else in nature. Upon digestion, a molecule of lactose yields one molecule of glucose and one of galactose. There is evidence that galactose plays an important role in the nervous system, particularly in the brain. Galactose is synthesized or converted from glucose in the alveoli of the mammary glands. It then is joined chemically with glucose to form lactose.

Lactose is commonly called milk sugar. It is less soluble than any of the other sugars that may be found in the digestive system, whether they be monosaccharides or disaccharides. It also must be digested before being absorbed. Because of its lower solubility level and its need for digestion, lactose requires a longer period of time to be absorbed than glucose or sucrose.

The percentage of sugar in the milk probably shows the least amount of variation among species compared to the variation found in fat and protein percentage. The milk of women has about 7.0 percent sugar while that of goats is about 4.3 percent. Mare's milk has about 6.0 percent sugar while that of the sows, ewes, and cows is about 5.0 percent.

Dairy Products

As was indicated earlier, over one-half of the milk produced in the United States is converted to some type of product other than fluid milk. Among the dairy products are included whole and nonfat dry milk, buttermilk, evaporated and condensed milk, whipping and coffee cream, flavored milk, ice cream, sherbet, yogurt, butter, cottage cheese, and cheddar cheese (see Fig. 15-2).

The fat, protein, and carbohydrate of milk can be separated to some extent in processing. The different dairy products contain varying amounts of these three substances. When whole unhomogenized milk is allowed to set and curdle, three fractions result. The cream rises to the top because it has a lower density or specific gravity than the other components. The protein called casein curdles and forms the curd or clabber; casein is the major protein in milk. The liquid portion that remains after milk has curdled is called whey. Lactose, minerals, B-vitamins, some of the protein, and other water-soluble materials are found in the whey. Whey contains about one percent protein.

A special kind of centrifuge, or cream separator, is used to remove the fat from whole milk. Butter and various kinds of cream are prepared from this portion. As marketed, butter must contain at least 80 percent fat. Butter is made by subjecting cream to a mechanical churning action. This causes the fat globules to cling together forming what we call butter. The amounts of fat in whipping cream, coffee cream and half-and-half are 40, 20, and 12 percent, respectively.

Considerable amounts of water may be removed from milk

Figure 15-2 Variety of dairy foods. Beginning at the lower right and reading clockwise: cream cheese, butter, sour cream, cottage cheese, pitcher of milk, Provolone cheese, and ice cream. Center: Swiss cheese atop Cheddar cheese, and to the left, Scamorze cheese and yogurt. (Courtesy United Dairy Industry Association, Rosemont, Illinois)

to form condensed or evaporated milk. The major difference between condensed and evaporated milk is that condensed milk contains additional sugar. If essentially all of the water is removed, whole or nonfat dry milk powder can be produced. Dry milk as well as evaporated and condensed milk may be reconstituted by the addition of water.

Several different types of cheese can be made from milk. Some are made from whole milk, some from low-fat milk, and some from skim milk. In the making of cheese, bacterial fermentation, treatment of enzymes, or a combination of both may be used. Cheddar or American cheese is composed mainly of the fat and protein fractions of the milk (see Table 15-2). Cottage cheese is made from skim milk and contains mainly protein.

Frozen desserts include ice cream, ice milk, sherbet, and frozen yogurt. Ice cream contains cream, milk, flavoring, added sugar, and corn syrup. Ice milk is made in essentially the same way as ice cream

TABLE 15-2. COMPOSITION OF SELECTED DAIRY PRODUCTS

Food	Percent Composition as Consumed					Percent of Total Solids		
	Water	Fat	Carbohydrate	Protein	Total Solids	Fat	Carbohydrate	Protein
Whole milk	87	4	5	3.5	13	31	38	27
Skim milk	90	Trace	5.5	4.0	10	Trace	55	40
Ice cream	63	11	21	4	37	30	57	11
Cottage cheese	80	Variable	3	14	20	Trace	15	70
Cheddar cheese	37	32	2	25	63	51	3	40
Butter	16	81	Trace	Trace	84	96	Trace	Trace
Yogurt	89	2	5	3	11	18	45	27
Sherbet	68	Trace	1	29	32	3	Trace	91
Evaporated milk	74	8	10	7	26	31	38	27
Condensed milk	26	8	56	8	74	11	76	11

Source: Adapted from USDA data.

but has a lower fat content. Sherbets are fruit-flavored products containing a very low amount of total milk solids.

A number of fermented milk products are produced which vary in fat content from that of whole milk to that of skim milk. These products include buttermilk, yogurt, sour cream, and sour cream dips. Each has its own procedure for developing the desired degree of fermentation and flavor.

THE EGG-LAYING CYCLE IN BIRDS

Female chickens and turkeys normally have only one ovary and one oviduct. Refer to Figure 15-3 for a diagram of the reproductive tract of hens. The term oviduct refers to the entire reproductive

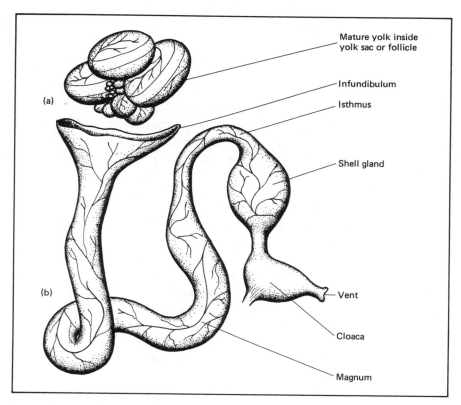

Figure 15-3 Ovary and oviduct of a chicken: (a) ovary, (b) oviduct.

tract in female poultry. In mammals, when this term is used, it refers only to that part of the tract from the ovary to the uterus. In a nonlaying hen the oviduct may be about 12 cm long. In laying hens it is much longer, up to about 80 cm. In laying turkey hens it would be several cm longer than in the chicken.

There are five parts or regions to the oviduct. The infundibulum lies near the ovary. It partially surrounds the ovary and functions in gathering the yolk when it is released from the ovary. The magnum of the oviduct corresponds to the ampulla of the Fallopian tube. The white of the egg, or albumen, is secreted by the walls of the magnum. The next part of the tract, the isthmus, secretes the shell membranes. The shell gland is next and its name suggests its function. The shell gland is sometimes referred to as the uterus. The fully formed egg is expelled through the vagina when it is layed. A more technical term for the egg-laying process is oviposition. The vagina of the oviduct is not an organ of copulation as it is in mammals. The male bird does not have a penis, so the vagina is not entered during copulation. The oviduct terminates at the cloaca. The cloaca is the common junction of the reproductive, urinary, and digestive tracts in birds.

Many of the same hormones that control reproduction in female mammals also control reproduction in female birds. Follicle-stimulating hormone (FSH) stimulates the growth and development of the follicles. At puberty the ovary contains many primary follicles as is the case in mammals. In birds, however, the ovicyte contains much more nutrient material than is found in mammalian ovicytes. At puberty the yolk-filled follicles are about 5 mm in diameter. After puberty, FSH causes the follicles to develop further, one at a time. Within about a week to ten days the first yolk and follicle will have increased to about 20 to 25 mm in diameter.

Estrogen is produced by the ovary. It stimulates the growth of the oviduct, causes the feathers and comb to take on the characteristic female appearance, and stimulates relaxation in the area of the pubic bones and vent in anticipation of oviposition, or egg laying.

Luteinizing hormone (LH) induces ovulation of the yolk from the follicle. The germ spot or true ovicyte is located on the outer surface of the yolk. As the yolk is picked up by the infundibulum it will be fertilized if the hen has been bred. The life of sperm inside the oviduct of the female is about 30 days or so. This means that a hen may lay fertile eggs for two or three weeks after she has been inseminated. In contrast to the mammalian ovary, no corpus luteum is formed after ovulation in birds.

As the yolk passes down the oviduct the other parts of the egg are deposited around the yolk. In about three hours in the magnum, the albumen, or white of the egg, is deposited. A little over one hour is required in the isthmus for the shell membranes to be secreted and deposited around the albumen. The longest time is spent in the shell gland, about 19 to 20 hours. The shell is composed of almost pure calcium carbonate which accounts for about 11 percent of the egg by weight.

Within a few minutes after leaving the shell gland, the egg will have passed through the vagina and the egg-laying cycle will have been completed. A hormone from the posterior lobe of the pituitary gland called vasotocin is responsible for oviposition, the expulsion of the egg. Vasotocin is very similar to the hormone oxytocin produced by mammals. From the time the egg leaves the ovary until it is layed, about 25 or 26 hours will have elapsed.

Ovulation in the chicken usually occurs about 30 minutes after laying the previous egg. Further, ovulation usually occurs during the morning daylight hours and very seldom occurs after two o'clock in the afternoon. Considering all of these time elements, a hen will usually lay an egg no less than 26 hours after she has laid the previous egg. An egg that is laid late in the afternoon will not be followed by ovulation until the next morning. This means that a hen may lay an egg each day for several days and then skip a day before she lays another. The number of eggs laid by a hen on consecutive days is known as a cycle or clutch. The shorter the egg-laying interval, the longer the clutch. The longer the clutch, the more eggs produced.

POULTRY PRODUCTS

Poultry Meat

The bulk of the edible parts of a chicken or turkey is made up of the skeletal muscles. Poultry meat differs from the meat of cattle, hogs, and sheep in several respects. The fat content of the meat from chickens is lower than that of the red meats. Most of the fat of chickens is located around the internal organs. Turkeys are usually fatter than chickens but not as fat as cattle, hogs, and sheep.

The pigmentation of poultry meat is different from that of red meat. Chicken fat carries more of a yellow color because of the presence of carotene. Some parts of the chicken carry red or dark pigments while others do not. This difference can be noticed when comparing breast meat with that of the leg. Breast meat carries very little pigment.

Because of the lower amount of fat, the meat of chickens usually has a higher percentage of protein than beef, pork, and lamb. While the figures vary greatly, the protein content of beef will usually be about 16 to 20 percent while that of chicken meat will be around 20 to 24 percent.

Protein quality of chicken meat is comparable to that of the red meats. All of the amino acids considered essential in human diets are present in more than ample amounts.

In terms of retail prices, chicken meat is usually a better buy than most of the cuts of red meat. For example at $2 per pound and 20 percent protein, beef protein from a sirloin steak would cost $10 per pound of protein. At $0.60 per pound and 24 percent protein, chicken would cost only $2.50 per pound of protein.

Dressed, ready-to-cook chickens, turkeys, ducks, geese, guinea fowl, and pigeons are quality graded according to specifications determined by the U.S. Department of Agriculture. Three grades are used: U.S. Grade A, U.S. Grade B, and U.S. Grade C.

Broiler production has come to be a rather efficient enterprise in terms of pounds of feed needed to produce a pound of bird. A 3-lb broiler, or fryer as they are sometimes called, can be produced on less than 6 lb of feed. This is a feed conversion ratio of less than 2 : 1. This compares to about a 3 : 1 feed conversion ratio of hogs and a 7 : 1 ratio for cattle.

The annual per capita consumption of poultry meat in the United States is about 52 or 53 lb. Most of this, about 43 lb, comes from chickens; approximately 9 lb comes from turkeys. All other kinds of poultry meat consumed amounts to less than one-third of a pound per person. The consumption of poultry meat usually increases when the consumption of red meat decreases; this has been the case during the past several years as noted in Table 15-1. As the consumption of red meat, particularly beef, increases, poultry meat consumption usually decreases. As mentioned earlier, the price per pound of ready-to-cook poultry meat is usually much lower than that of red meat. When the proportion of consumers' income that is available for food decreases, they tend to purchase more of the lower-priced meat and less of the higher.

Eggs

An egg is the reproductive cell of a female. The unfertilized egg contains one-half the genetic makeup of a body cell of the female parent plus varying amounts of nutrients to help the young embryo to develop until it is provided with another source of nutrition. In order for the embryo to develop, it must have been fertilized by the male reproductive cell, the sperm, and properly incubated. This description of an egg applies to essentially any group of animals whether it be mammals, reptiles, insects, fish, or birds.

Compared to mammalian eggs, avian eggs are generally much larger relative to the mature size of the adult. The embryo develops inside the egg shell after the egg is laid rather than inside the uterus in the case of mammals. Bird eggs also contain a much greater amount of nutrients—enough to nourish the embryo until it hatches. The extra amount of food material is what accounts for the greater size of the eggs of birds compared to mammalian eggs.

The parts of an egg include the shell, shell membranes, yolk, and albumen. The albumen is usually referred to as the white of the egg because it turns white when the egg is cooked. In fact, the word albumen comes from a Latin word that means "white." The shell membranes are not so obvious in a fresh egg; they normally remain with the shell when the egg is broken. When a boiled egg is peeled, however, the shell membranes can be seen rather well. They are made up of a fibrous protein whose weight is almost negligible when compared to the weight of the whole egg. The parts of a chicken egg can be seen in Figure 15-4.

The shell is composed mainly of calcium carbonate and accounts for a little over 10 percent of the total weight of the egg. The food value of an egg is found in the yolk and albumen. The albumen is composed of about 89 percent water and 10 percent

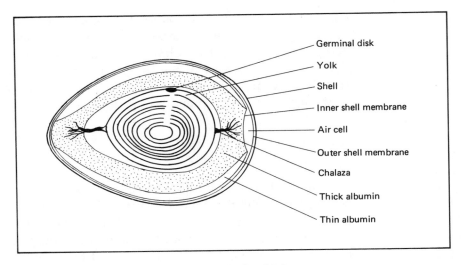

Figure 15-4   Parts of a chicken egg.

protein. The other 1 percent is made up of a number of organic and inorganic substances. The albumen accounts for almost 60 percent of the weight of an egg.

About 30 percent of the weight of an egg is found in the yolk. The yolk contains protein and fat primarily, but also contains some vitamins, minerals, fatty acids, and cholesterol. Because of their cholesterol content, eggs have been the subject of some controversy with regard to the possible relationship between cholesterol and heart disease. This subject is presented in more detail in the next section of this chapter.

The protein in eggs is considered to be of the highest quality among foods commonly consumed by humans. The essential amino acids are present in more than adequate amounts and make up a larger percentage of the total protein compared to other foods. The annual consumption of eggs in the United States is about 275, or 23 dozens, per person.

Eggs are quality graded according to soundness of the shell, size of the air cell (the fresher the egg, the smaller the air cell), and whether the white and yolk are free of blood spots and other defects. Grading is done by subjecting the eggs to a process called candling, in which they are passed over a lighted background so that

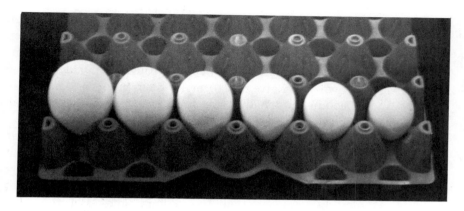

Figure 15-5 Variation in size of chicken eggs. The sizes from left to right are Jumbo, Extra large, Large, Medium, Small, and Peewee.

the size of the air cell and interior quality can be determined. The quality grades are USDA Grades AA, A, B, and C.

Eggs are classed according to size by weight in ounces per dozen. The size and weight classes are Jumbo (30 oz per dozen minimum), Extra Large (27 oz), Large (24 oz), Medium (21 oz), Small (18 oz), and Peewee (15 oz). See Figure 15-5. The average large size egg weighs approximately 1 1/2 ounces.

FATS AND CHOLESTEROL IN HUMAN DIETS

Interest in cholesterol and animal fats in human nutrition has increased during the past 25 to 30 years because of an apparent relationship between their consumption and the incidence of heart disease. Over one-half of the deaths of people in the United States in recent years has been due to heart disease. One type of heart disease is called atherosclerosis, which refers to an accumulation of substances in the inner linings of the arteries. Some of these substances include proteins, fats, iron and other minerals, phospholipids, and cholesterol.

The buildup of substances in the wall of the arteries continues over the years so that, as a person becomes older, the deposits become thicker. At least two effects result from this buildup. First, the inside diameter of the arteries becomes smaller, resulting in a smaller overall capacity to the circulatory system. Second, the walls of the arteries become rough and calcium deposits are laid down in an attempt to make them smooth. The walls of the arteries become hardened, giving rise to the common term for atherosclerosis: "hardening of the arteries."

Some research studies have indicated that people with high blood cholesterol levels stand a greater risk of having atherosclerosis. Because of this and the fact that cholesterol is one of the several substances found in the walls of the arteries, there has been a great public concern that the dietary intake of this substance should be reduced. Because eggs, milk, and meat are major sources of cholesterol, there are those who advise against the consumption of these foods. In addition, cholesterol tends to be associated with the fat in animal products. For example, skim milk contains very little cholesterol while whole milk (with between 3 and 4 percent butterfat) contains 60 to 70 mg per pint. Because of this relationship between cholesterol and animal fats, some well-meaning people have also indicted animal fats as a major contributor to heart disease.

In view of this controversy, it should be kept in mind that cholesterol is involved in the synthesis of several important and vital body substances. A large part of the cholesterol in the body is converted to bile in the liver. The sex hormones estrogen, progesterone, and testosterone are synthesized from cholesterol. Hormones from the cortex of the adrenal glands are likewise synthesized from cholesterol. The body can also convert a derivative of cholesterol to vitamin D with the help of ultraviolet light from the sun. It should be apparent that without cholesterol in the body it would be extremely difficult, if not impossible, to survive. The healthy body maintains a normal level of cholesterol in the blood by synthesizing it in the liver. The more cholesterol consumed in the diet, the less is synthesized. If too much is consumed or synthesized, the excess amount is degraded or excreted to keep the amount in the blood between normal limits. The liver synthesizes about twice as much cholesterol as is consumed in the average daily diet.

No attempt will be made to analyze the many hundreds of

volumes of research data that exist concerning cholesterol and its relationship to human nutrition and heart disease. Many of the experiments were well designed; many probably were not. Many studies have produced results that contradict the results of other studies. Much of the research was conducted or financed by organizations and institutions with a vested interest in one side or the other of the controversy. To say the least, the problem is not a simple one and no single cause-and-effect relationship has yet been established.

Many scientists are now beginning to conclude that the indictment of animal fats and cholesterol as major causes of heart disease has been premature and very likely incorrect. Many other causes and predisposing factors are also associated with heart disease. Among those factors are heredity, social and business stresses, drugs, alcoholic beverages, smoking, obesity, and physical inactivity.

Several other interesting factors concerning cholesterol and animal fats tend to cast additional doubts upon the idea that these substances are as dangerous as some people would like us to believe. The incidence of coronary heart disease in the United States has approximately doubled during the past 70 years. During that period of time, the per capita consumption of animal fats has decreased to one-fourth the original amount while the consumption of plant fats has more than tripled.

The United States leads the world in number of deaths due to coronary heart disease, but we rank about fifteenth or sixteenth in per capita consumption of milk fat. The people of Canada, New Zealand, the Netherlands, and some of the Scandinavian countries consume very liberal amounts of animal products, but the life expectancy in those countries is greater than in the United States.

Herbivorous animals consume no animal fats or cholesterol, yet their blood cholesterol levels are about the same as carnivorous animals. The reason for this goes back to the function of the liver in synthesizing this vital substance. Herbivorous animals have been known to suffer from atherosclerosis. The disease has been reported in elephants, deer, sheep, cattle, African buffaloes, and goats.

The blood cholesterol levels of pregnant animals are normally higher than nonpregnant animals. Eggs are high in cholesterol, yet they were designed to provide nutrition for a developing embryo.

Milk contains cholesterol, but the natural function of milk is to provide nutrition for the newborn.

With regard to just these few examples, it does not seem logical to blame the consumption of animal fats and cholesterol for the increased incidence of coronary heart disease in the United States.

STUDY QUESTIONS AND EXERCISES

1. Define each of the following terms:

   albumen (albumin)   hardening of the arteries
   atherosclerosis   homogenization
   broiler   ice cream
   butter   ice milk
   butterfat   lactation
   buttermilk   LH
   candling   magnum
   casein   pasteurization
   cholesterol   prolactin
   cloaca   udder
   cream   vasotocin
   curd   whey
   dry   yogurt
   egg   yolk
   emulsion

2. Describe and sketch the basic structure of a mammary gland of a cow.

3. What is the relationship between the number of mammary glands present in a particular species and the number of offspring they normally have?

4. How is milk made?

5. Explain how milk is removed from the mammary gland once it has been secreted.

6. About what percentage of the milk used for human consumption comes from cattle?

7. What is the approximate annual consumption of milk per person in the United States?

8. What has been the trend in milk consumption over the past 25 years?

9. Distinguish between Grade A and Grade B milk and Class I and Class II milk.

10. How does the newer method of pasteurization differ from the older method?

11. Compare the nutrient content of whole milk to that of meat.

12. Name a mineral for which milk is a poor source and one for which it is a good source.

13. Name a vitamin for which milk is a poor source and one for which it is a good source.

14. Compare the percentage of protein in the milk of farm mammals and women.

15. Compare the percentages of fat and sugar in the milk of farm mammals and women.

16. Compare the solubility, sweetness, and digestibility of lactose and sucrose.

17. Compare the percentages of fat, sugar, and protein in the various common dairy product foods.

18. What is the difference between condensed and evaporated milk?

19. Name several dairy products that are produced with the aid of fermentation.

20. Describe the process of egg formation in the oviduct of a laying hen.

21. Explain why a hen does not usually lay an egg every day during the laying season.

22. How is it possible for a hen to be able to lay fertile eggs for several weeks without being exposed to a cock, or rooster?

23. Compare the characteristics of poultry meat and the red meats.

24. Compare the costs of protein from chicken to that of pork and beef at today's prices.

25. What are the quality grades for dressed poultry meat?

26. Compare the feed conversion ratios of broilers to that of hogs and beef cattle.

27. Compare the annual consumption of poultry meat to that of milk, beef, and pork.

28. Why are chicken eggs so much larger than mammalian eggs?

29. Compare the nutrient composition of the egg yolk to that of the albumen.

30. What is the difference in weight per dozen between any two size classes of chicken eggs?

31. How is the size of the air cell related to egg quality?

32. What are the USDA quality grades for eggs?

33. Describe the effects that result from the buildup of materials in the walls of the arteries.

34. What is the apparent relationship between saturation of fats and heart disease?

35. List some of the nondietary factors that may contribute to heart disease.

36. What important functions does cholesterol have in the body?

37. What are the sources of blood cholesterol?

38. Which common foods have the highest cholesterol content?

39. What has happened to the incidence of heart disease in the past half century or so?

40. What has happened to the consumption figures for animal fats during the past 50 to 60 years?

41. Compare the cholesterol levels in herbivores and carnivores.

16

Animal Health and Disease

Health is the state of well-being of
an animal in which all body parts are functioning normally. Disease
refers to any disturbance in the normal structure or function of any
body part. The definition of health is positive, indicating a good
situation. Disease suggests a negative condition—a situation that is
not good. An animal with a disease can be described as one that is in
poor health.

A knowledge of diseases and their control is important for a
number of reasons. Diseased animals do not produce at their ultimate
capacity. If the disease is severe, it may result in the complete loss of
animals through death or condemnation. Any of these situations results in an economic loss to the livestock producer. Moreover, since
livestock diseases can be transmitted to humans, there is a connection between the state of health of domestic animals and human
or public health.

CHARACTERISTICS OF HEALTH AND DISEASE

There are varying degrees of severity of disease. Some diseases may
be rather mild where others may be severe and even cause death. A
disease is said to be acute if it occurs quickly and runs its course

within a few days. A very acute disease may appear and result in the death of an animal before any symptoms can be noticed. A chronic disease is one that shows symptoms very slowly and has its effect over a much longer period of time. Tuberculosis and certain fungal infections are examples of chronic diseases. Anthrax, milk fever, and various poisons are usually acute.

Diseases are usually catergorized as infectious, noninfectious, and parasitic. Infectious diseases include those caused by microorganisms such as bacteria, viruses, and protozoa. They involve the invasion of the host by another organism. Parasitic diseases are technically infectious also, because they involve the invasion of the host by a foriegn organism. However, the organisms that are usually considered parasites include worms and arthropods and are generally not microscopic. Noninfectious diseases include disorders caused by factors other than organisms. Examples include mechanical problems such as wounds and broken bones, digestive disturbances such as bloat, nutritional deficiencies such as rickets and night blindness, poisoning due to toxic plants or to chemicals such as pesticides, as well as genetic and metabolic disorders.

Distinguishing between Healthy and Diseased Animals

Before we get into a discussion of specific organisms and how to prevent diseases, we need to consider some of the characteristics of the normal or healthy animal. Knowing some of the normal characteristics and behavior patterns of domestic animals makes it easier to detect certain disease problems when they appear. When an animal contracts a disease, its normal behavioral routine is usually altered. Certain physiological characteristics are also affected by the presence of disease.

Healthy animals tend to be content, alert, and bright-eyed. The saying that someone is "bright-eyed and bushy-tailed" is a reference to looking and being healthy. If an animal does not seem to be annoyed easily, stands with its head and neck down, or appears to have a tired or lazy look, it may be affected by some disease. A dullness or lack of color in the eyes may indicate a problem. See Fig. 16-1.

Animals should show normal eating habits; cattle and sheep should be observed chewing the cud. The appearance of the feces and

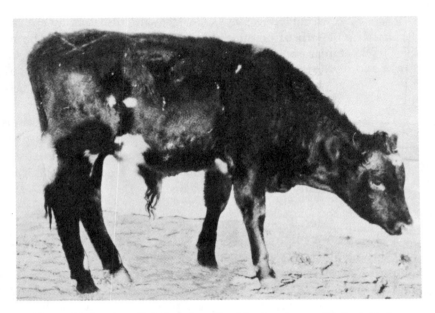

Figure 16-1   A calf infected with shipping fever. (United States Department of Agriculture)

urine are often indicators of the state of health. The hair coat should be sleek and appropriate for the season. If cattle and horses fail to shed their winter coats in the spring and early summer, this could be an indication of some internal parasite.

Deviations from normal breathing rates, pulse rates, and body temperature may be an indication of poor health. Fever has been recognized as a sign of disease for several thousand years. A number of kinds of infection may cause a rise in body temperature by interfering with the temperature-regulating mechanism of the body.

The Comfort Zone

Changes in body temperatures can be brought about by exposure to relatively high or low temperatures for prolonged periods of time. The range of ambient temperatures between which body temperatures can be maintained is called the comfort zone. When the environmental temperature falls below about 32°F (0°C) the thermoregulatory mechanism in European cattle must become active to keep the body temperature from falling. If the environmental temperature rises above about 60°F (16°C), the thermoregulatory mechanism must function to keep the body temperature from rising.

If the environmental temperature remains in the vicinity of 80°F (27°C) for an extended period of time, the temperature-regulating mechanisms in European cattle cannot keep their body temperature down to normal and a fever will develop. The comfort zone for Zebu cattle is in the range of about 50° to 80°F. It is interesting to note that for humans the comfort zone is about 65° to 80°F.

Transmission of Infectious Diseases

The transmission of disease can be thought of as a chain with three links (see Fig. 16-2). The first link is the source or reservoir for the disease; this is usually a diseased animal. The second link is the mode of transmission. The third is a susceptible animal.

Several modes of transmission of disease exist. Diseases may be transmitted by direct or indirect contact, or by a vehicle or vector. Animals may come in contact directly with diseased animals or indirectly through contact with feces, urine, or an aborted fetus.

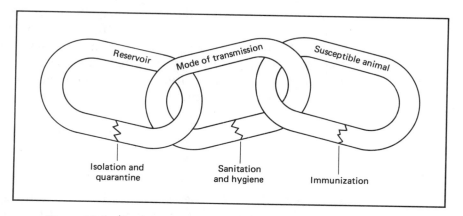

Figure 16-2   Three links in the chain of disease transmission. Included are some factors that can cause any one of the links to be broken which in turn disrupt the entire chain of transmission.

Organisms carried by way of fine droplets of moisture as from coughing or sneezing can be thought of as indirect contact. Brucellosis and vibrosis can be transmitted by direct contact.

Various vehicles can be involved in the transmission of disease organisms. These involve food, water, and infected medical instruments. Carrier animals that may not be susceptible to a certain disease can transmit the organisms. Trucks, trailers, and portable handling equipment can become vehicles for disease transmission. Examples of waterborne diseases, for example, are leptospirosis and anthrax.

Insects and arachnids may act as vectors by carrying disease organisms in or on their bodies. Face flies can carry the organism that causes pinkeye; certain ticks carry the organism for Rocky Mountain spotted fever; encephalitis is transmitted by several types of mosquitoes. In some cases the vector acts as a passive carrier; that is, the disease organism is carried in or on the body of the arthropod but does not depend upon it for survival. Such is the case of the pinkeye organisms carried by face flies. This is not the situation with malaria in humans, for example. The malaria organism depends upon the vector for a part of its life cycle. In other words, the vector acts as an intermediate host for the disease organism. Without the intermediate host, the organism could not survive.

In order for the disease organisms to cause disease there must obviously be a susceptible animal. If there is no susceptible animal, the chain of disease transmission is broken. Assuming that a susceptible animal is available, the organism can enter the body of the animal by way of a number of openings. Probably the most common entryways are the mouth and nose. Other openings that can serve as routes of entry are the eyes, the opening to the reproductive tract, and wounds. Entrance may be gained by openings made by biting arthropods and contaminated medical equipment such as vaccinating needles and castrating knives.

VETERINARY PREVENTIVE MEDICINE

The prevention of diseases caused by infectious agents involves actions that are taken to break one or more of the three links in the chain of disease transmission described earlier. It is more economical and certainly much safer for domestic animals and humans to prevent the occurrence of a disease rather than to try to cure the disease once it has been contracted. Breaking the first two links of the chain involves principles of sanitation and containment of the disease organisms. Breaking the third link involves the defense mechanisms inherent to the body of the animal, and the protection of the animal by the use of immunizations.

At this point, a word should be said about the role and importance of veterinarians. Veterinarians are engaged in activities related to the study, prevention, and control of animal diseases. In other words, they are concerned with maintaining animal health. They cannot do this job alone. They must have the help of all others concerned including farmers, ranchers, physicians, biologists, and any number of other specialists upon whom they can call.

Too often a farmer may observe signs of poor health in some of his or her livestock and will not contact the veterinarian soon enough. This tendency to delay may be due in part to the hope that the animal is not really sick enough to need the services of a veterinarian, or the farmer may be trying to save the expense of the service call. Very often, by the time a veterinarian is finally called, the disease has progressed to the state that the animal has a very poor chance of recovering. It behooves the livestock farmer to know enough about disease problems to be able to recognize whether or not it is desirable to call a veterinarian.

Principles of Disease Prevention

One important principle of disease prevention involves simple isolation or quarantine. When an animal is known or suspected of having a contagious disease, it should be separated from the rest of the herd until the disease is isolated and controlled. Quarantine may be as simple as putting an animal in a separate building or pasture away from other animals, or it may take a more drastic form in that a whole farm may be put under official legal quarantine. For example, some states have laws that require whole herds to be quarantined when an animal in that herd shows a positive test for brucellosis.

When newly purchased animals are brought into a herd, it is wise to keep them in isolation for two or three months until it is known that they are not carrying any disease organisms. When stock are returned from shows they should be held in isolation as if they were newly purchased. Examinations of the isolated animals should be made by a veterinarian.

Sound sanitation practices should also be followed to prevent diseases from being transmitted. Keeping animals on clean ground is important if it is at all possible. This will involve rotating pastures and feedlots, changing and cleaning stalls, and properly disposing of manure. Dead animals should be disposed of properly. The best method of disposal is to have the larger animals picked up by a dead-animal truck. It is not wise to try to burn dead animals or to feed them to other animals. It is amazing how much wood is required to completely burn the carcass of even a 30- or 40-lb pig.

Sanitation procedures often involve the use of disinfectants and antiseptics. Disinfectants are chemical or nonchemical agents used to kill disease organisms before they have a chance to cause infections. These agents may be used on the animal itself or they may be used in the cleaning of utensils and objects with which animals may come in contact. Antiseptics are chemical substances used to kill organisms to keep them from entering the body tissue. The definitions of the two terms are not very different. Some of the chemicals used as disinfectants or antiseptics include alcohol, tincture of iodine, phenol, mercurochrome, and cresol. Nonchemical disinfectants may include extremes of either hot or cold such as flame, steam, boiling water, and freezing temperatures. Ultraviolet light from the sun or other source is also an excellent disinfectant.

The Body's Defense Mechanisms

The body's first line of defense against disease is epithelial tissue. Epithelium includes the skin and mucous membranes. The latter includes the tissue that lines the inside of the mouth, nasal cavity, and other body openings. It is very difficult for bacteria to penetrate these tissues unless there is a break in them. When epithelial tissue has become broken by surgery, wounds, or other causes, it is highly susceptible to entrance by disease organisms.

Some epithelium produces secretions that act as antiseptics. An antispetic is a substance that interferes with the normal growth and development of microorganisms, particularly disease organisms. Alcohol, iodine, and mercurochrome are widely used antiseptics. Natural antiseptics of the body include lysozyme found in certain body secretions such as tears. The hydrochloric acid produced in the stomach exhibits antiseptic qualities by destroying disease-producing organisms that may be swallowed in food and water.

If disease-causing organisms should get through the first line of defense, any one or more of several other mechanisms may be called upon to defend the body against the invader. One of these mechanisms is the inflammation reaction. However, it should be emphasized that the inflammation reaction can be brought about by causes other than infection. Wounds, burns, chemicals, and other factors that can cause tissue damage may cause inflammation.

The four symptoms usually associated with inflammation are swelling, redness, fever, and pain. The redness is due to an increased blood supply to the damaged tissue. The increased blood pressure in the vessels causes fluids to enter the tissue outside the blood vessels in the damaged area, which in turn causes swelling. The fluid that leaks from the blood vessels is called exudate. The increased blood pressure and the damage to the tissues results in an increase in the temperature in the damaged or infected area. Pain is an obvious consequence of damage to tissue.

An increase in the overall body temperature can come about due to infection. This is called the febrile reaction. Toxins produced by microorganisms can affect the temperature regulatory mechanism of the brain. Results of some recent research indicate that the fever itself has a healing effect upon the body.

The white blood cells are also defenders against disease. One group of white cells, called neutrophils, are phagocytic cells. They engulf invading organisms and attempt to destroy them. Neutrophils greatly increase in number at the site of an infection. Many neutrophils are killed by the disease organisms. The dead white cells and other materials at the site are called pus. For this reason, neutrophils are sometimes called pus cells.

Another group of white blood cells are the lymphocytes. The cells are produced by lymph nodes, spleen, and other lymphoid tissue. They are associated in some way with the production of antibodies which help to combat disease organisms. The lymph nodes themselves make up a filtering system in the body to catch microorganisms. They are a part of the lymphatic circulation system. The liver and spleen also play a role in the filtering mechanism of the body.

Immunity is that characteristic of an animal that allows it to resist being overcome by a disease. Many animals bear a natural immunity to certain diseases. Cattle have a natural immunity to malaria and measles organisms. Humans are resistant to Texas cattle fever. Much natural immunity is due to the presence of immune substances in the body fluids called antibodies. Some of these antibodies are produced genetically while others are developed by the body after the disease organism enters.

Principles of Immunization

A substance that can enter the body and stimulate the production of antibodies is called an antigen. In the context of disease prevention this could be a disease-producing organism. The antibody in turn has a detrimental effect upon the organism that stimulated its production. This reaction is referred to as the antibody-antigen reaction. It is the basis for our discussion of immunization.

Immunity may be inherent or acquired. Inherent immunity is the kind that is due to the particular genetic makeup of the individual or species. Acquired immunity is the kind obtained from the introduction of antigens or antibodies from a source outside the body. Animals can acquire immunity naturally or artificially; the animal may plan an active role or a passive role in the development of the immunity; and immunity may last for only a few weeks or it may last for several years.

When an animal produces its own antibodies against a disease, it is said to play an active role in its immunity. Active immunity is acquired by the introduction of antigens into the system. The antigens in turn stimulate the production of antibodies which provide a certain level of immunity to future introductions of that particular antigen. The antigens may be introduced naturally if an animal contracts a disease. This is not the recommended way to develop immunity because it is difficult to control the organisms when introduced in this way.

The better way to develop active immunity in an animal is through the introduction of a vaccine. A vaccine is an antigen preparation that contains either the infectious agent itself or its metabolic products. If the vaccine contains dead organisms, it is referred to as killed vaccine. The dead organisms are able to induce the production of antibodies after they have been introduced but are harmless in themselves. These preparations are also called bacterins. An example is blackleg bacterin.

Another type of vaccine contains live organisms that have been modified so they are not as harmful as the untreated organisms. They are called modified live vaccines or attenuated vaccines. Some types of rabies and brucellosis vaccines are produced in this way. Still another type of vaccine contains only toxins produced by the infectious organisms rather than the actual pathogens themselves. One of the best-known examples of this type of vaccine is the tetanus toxoid vaccine. Characteristics of acquired immunity are illustrated in Figure 16-3.

The immunity produced by vaccination may last for months or years depending upon the particular type of organism. Immunizations of this kind are performed to prevent animals from being overcome by a disease should they be exposed.

A second major type of acquired immunity is called passive immunity. With this kind of immunity, the animal does not play an active role in the development of its immunity. Antibodies for a given disease are produced outside the body and injected into the animal. Disease antigens may be injected into a laboratory animal or any animal that does not show a high degree of susceptibility to the disease. The blood of the laboratory animal will produce antibodies

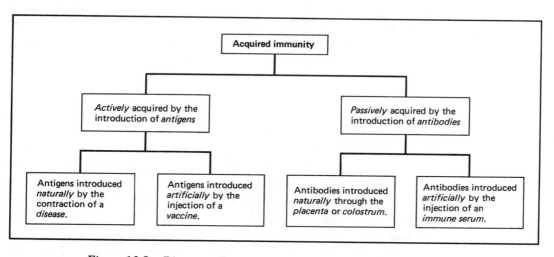

Figure 16-3   Diagram illustrating the several types of acquired immunity.

against the disease. Some of that blood can be removed and allowed to clot. The serum can be separated and prepared for injection as an antiserum. Swine erysipelas and scarlet fever in humans are two diseases for which antiserums have been prepared.

Passive immunity can be acquired naturally in at least two ways. In some species immune dams can transmit antibodies through the placental membranes to their fetuses. The young are born with some degree of immunity. Antibodies can also be secreted in the first milk or colostrum of immune mothers. For this reason it is important that newborn animals obtain the first milk from their mothers. Colostrum also cantains a high percentage of protein and vitamin A which are beneficial to the newborn animal.

As a prevention against disease, the type of immunity developed from the injection of antiserums is relatively short lived, sometimes only a few weeks. With some diseases, no effective vaccine has been developed, so antiserums may be the only means of producing immunity. Because antiserums provide an immediate supply of antibodies, they may be used in the treatment of an animal that has contracted a disease.

MICROORGANISMS

Disease-producing organisms that are microscopic in size include bacteria, rickettsia, viruses, and protozoa. Although not all fungi are microscopic, those that are responsible for many animal diseases are; therefore, they are included in this section.

Bacteria

Bacteria are one-celled organisms varying is size from about 0.5 to 10 μ in diameter. In the two-kingdom system of classifying organisms they were placed in the plant kingdom. In the multikingdom system of classification sometimes used today, bacteria are not included in the plant kingdom, and neither are the fungi. Bacteria do not contain chlorophyll.

Bacteria play many essential roles in the ecology of plant and animal life. Bacteria found in the root nodules of legume plants help in the manufacture of protein. Soil bacteria help in the process of decomposing organic matter and converting it to nutrients for plant growth. Certain kinds of bacteria aid in the production and processing of some foods through the fermentation process. Cheese, buttermilk, sauerkraut, and yogurt are made with the help of bacteria. The preservation and storage of silage depends upon the action of certain bacteria. Other kinds of bacteria found in the rumen of cattle and sheep aid in the proper digestion and metabolism of feeds.

The cellular structure of bacteria is in some ways similiar to that of other cells. But there are some contrasts, also. Practically all bacteria have a cell wall similar to that of plant cells. Inside the cell wall is found a cell membrane. Some bacteria are further covered with a soft capsule that provides added protection. Inside the cell membrane is found cytoplasm similar to that of other cells. The nucleus is somewhat different in that there is no nuclear membrane surrounding the single unit of DNA.

Bacteria may be adversely affected by such factors as heat, sunlight, ultraviolet light, antiseptic chemicals, and antibiotics. Antiseptics may either kill the cells or merely inhibit their growth. Antibiotics such as penicillin kill bacteria by interfering with the function of the cell wall. Antibiotics do not affect animal cells or viruses because they do not have a cell wall.

Most bacteria reproduce asexually by a process called fission which is similar to mitotic division. Some bacteria produce spores which are quite resistant to destruction and may live for years. These spores ultimately develop into bacteria. Bacteria in general reproduce very rapidly, being able to double their numbers in as little as 15 or 20 minutes.

Bacteria can be classified according to their need for atmospheric oxygen. Aerobic bacteria need atmospheric oxygen in order to grow. They will die without it. Anaerobic bacteria do not need atmospheric oxygen because they are able to produce their own. Exposure of anaerobic bacteria to an outside source of oxygen will inhibit their growth. The bacterium that cause tetanus is an anaerobe.

Bacteria are also categorized according to their general shape. These include the bacilli (rod-shaped), cocci (dot-shaped), spirochetes (spiral-shaped), and the vibrio bacteria (comma-shaped).

The bacilli are rod-shaped cells that may appear singly, in pairs, or in chains. The rods may be long or short. Many disease organisms fall in the bacillus group. Some examples are brucellosis, tuberculosis, tetanus, and anthrax. Some authors classify vibrio bacteria with the bacilli. They are curved or comma-shaped rods. A venereal disease of cattle called vibriosis is caused by this type of organism. Human cholera is also caused by a vibrio type of bacterium.

Spherical or dot-shaped bacteria are called cocci. They may grow in pairs in which case they are called diplococci. Clusters of cocci are referred to as staphylococci. Streptococcus bacteria are those that grow in chains. Spirilla, or spirochetes, are spiral-shaped bacteria. Leptospirosis is an example of a disease caused by this kind of bacterium.

The following diseases are but a few examples of those caused by bacteria. No attempt is being made to make the reader an expert in the diagnosis and treatment of these diseases. A brief discription is given to help develop an appreciation for the problems of bacterial diseases. If animals are suspected of harboring any diseases it is wise to consult a veterinarian.

1. **Anthrax** is caused by a spore-forming bacillus. Nearly all farm mammals are susceptible. Swine, poultry, and humans show considerable resistance to the disease. Anthrax is one of the oldest known diseases and the first for which an effective vaccine was developed. It is a highly fatal, infectious disease that affects the blood as well as tissues in general. The spores are highly resistant to the enviroment and may live in the soil for many years.

2. **Blackleg** is also caused by a spore-forming bacillus. The spores are highly resistant to destruction by heat, cold, and disinfectants, and may live in the soil for years. The blackleg organism is an example of an anaerobic bacterium. The disease affects cattle between 6 and 18 months of age; sheep are sometimes affected. Vaccination can provide an effective protection against the disease. The organism causes a gaseous swelling in the muscles, usually in the hind legs. See Fig. 16-4.

3. **Brucellosis** is caused by a bacillus and is also known as Bang's disease. Three species of the organism exist affecting mainly cattle, swine, and goats. When humans are affected by the disease it is called undulant fever. It is an infectious and widespread disease that affects the reproductive organs and causes abortions. It does not seem to be

Figure 16-4   Young animals (6 months to 4 years old) can be affected with the fatal backleg disease. (United States Department of Agriculture)

transmitted by copulation in cattle. No practical effective treatment is known. Control measures involve the testing of animals and the slaughter of those that show a positive reaction. An effective vaccine is available. See Fig. 16-5.

Figure 16-5   Many counties require brucellosis blood testing and disposing of reactors to control the disease. (United States Department of Agriculture)

4. **Erysipelas** organisms can survive for years in the soil. An outbreak of the disease can occur where seemingly it has never occurred before. It is a bacillus that particularly affects hogs and turkeys. A vaccine is available that provides fairly good protection but it is not completely effective. The disease apparently affects the joints and causes difficulty in breathing. Because skin lesions occur in some animals erisypelas is sometimes called diamond skin disease.

5. **Leptospirosis** is caused by a spirochete. The disease can cause destruction (hemolysis) of red blood cells and may cause abortion. The organisms affect the kidneys and tend to collect in the urine. Transmission of the disease usually involves contact with contaminated urine or water. It is essentially a waterborne disease. The disease can be transmitted to humans. An effective vaccine is available for domestic animals. See Fig. 16-6.

6. **Tetanus,** or lockjaw, affects the nervous system in such a way that the sick animal cannot control certain muscles including those that control the jaws. The tetanus bacillus is an anaerobe—it cannot grow in the presence of oxygen. Immunity can be induced

Figure 16-6   A cow and her aborted fetus, caused by leptospirosis. (United States Department of Agriculture)

by the injection of a toxoid vaccine. Humans and horses are quite susceptible, but the disease affects essentially all mammals.

7. **Tuberculosis** affects practically all birds and mammals, including humans. The organism is a bacillus that primarily affects the lungs in domestic mammals. In poultry, the disease affects mainly the liver and spleen. In humans several different tissues may be affected including the bones, joints, skin, nervous system, and lymph nodes. Humans and cattle are the main reservoirs of the disease. Control in livestock involves testing and slaughter of reactors. An effective vaccine is available; however, vaccination causes the diagnostic test to be invalid. For this reason, vaccination is seldom practiced.

8. **Vibriosis** of cattle is caused by the curved rod type of bacterium. The organism is transmitted by sexual contact, so it is classified as a veneral disease of cattle. Humans may obtain the organism in milk from infected cows or by contact with slaughtered animals. Vaccines may be used when the incidence of the disease is high. Control measures involve the use of artificial insemination using semen treated with antibiotics.

Viruses

The usual scheme of classifying living things into various kingdoms does not include the viruses. They are not cells and they cannot live outside of other living tissues. Viruses are so small that they generally cannot be seen through a light microscope. They must be observed using an electron microscope.

These submicroscopic agents are composed of a protein capsule containing nucleic acids. Some viruses contain DNA while others contain RNA. Viruses attach themselves to the outside of a cell and inject their nucleic acids into the cytoplasm of the cell. Many new viruses are formed inside the cell and then released to attack other cells. The host cell may be injured by the ordeal or even killed.

Antibiotics as well as many other drugs used to combat bacterial diseases are ineffective in combatting viruses. Since viruses are usually found inside of living cells it is very difficult to kill the viruses without also causing damage to the host cell. In dealing with viral diseases, prevention of the disease becomes the key factor rather than treatment.

Following are a few examples of viral diseases and some of their characteristics. Some of these diseases are not the problem that they once were. Examples of viral diseases in humans include chicken pox, the common cold, influenza, measles, mumps, poliomyelitis, and yellow fever.

1. Rabies is a disease of the nervous system that can lead to death if left untreated. The word rabies comes from a Latin word meaning "fury" or "rage." Rabid dogs or wild animals will often attack other animals and humans. The disease affects the muscles in the region of the pharynx such that rabid animals have difficulty swallowing. Because they cannot swallow they will not drink, so the disease is often called hydrophobia which means fear of water. All dogs and cats should be vaccinated. If pets or humans are bitten by rabid dogs or wild animals, a physician or veterinarian should be contacted immediately. All mammals are probably susceptible to the disease, including humans. The most common vectors of rabies are dogs, cats, skunks, foxes, wolves, raccoons, and bats. The virus is carried in the saliva and is transmitted by a bite or by the saliva coming in contact with a wound.

2. Foot-and-Mouth-Disease (FMD) is caused by one of the smallest viruses known, even smaller than the polio virus of humans. The virus affects the blood, lymph nodes, and bone marrow. Swine and domestic and wild ruminants are susceptible. Foot-and-mouth disease can be transmitted to humans. The virus can be carried in meat and milk products, on clothing, on the bodies of birds and mammals where it can live for several months.

3. Hog Cholera is of very little importance in the United States. In fact, it has been essentially eradicted. No vaccinations are being performed and all vaccines are being destroyed. The virus is transmitted by direct contact between pigs and by feeding uncooked pork. This disease should not be confused with human cholera which is caused by a bacterium. See Fig. 16-7.

Figure 16-7 Listless victims of hog cholera. (United States Department of Agriculture)

4. **Equine Encephalomyelitis,** meaning inflammation of the brain, affects horses and mules. It is also known as sleeping sickness, brain fever, staggers, and blind staggers. Outbreaks of the disease were common in the 1930s and 1940s when horse and mule numbers were relatively large. The incidence of the disease dwindled considerably after that until 1971 when an outbreak occurred in the southwestern United States. The disease can affect other mammals, including humans, as well as birds. The organism is transmitted by blood-sucking arthropods. Mosquitoes are probably the most common vectors. Effective vaccines are available for the various strains of viruses that exist. See Fig. 16-8.

5. **Infectious Bovine Rhinotracheitis** (IBR), commonly called red nose, is a highly contagious disease affecting the upper respiratory tract in cattle. In can also cause abortions and stillbirths, and encephalitis in young calves. It also causes infections of the vagina and vulva of cows and sometimes of the penis in bulls. A modified live-virus vaccine against IBR is available for nonpregnant cattle.

6. **Bovine Viral Diarrhea** (BVD) is a disease that affects the digestive tract of cattle and can also cause abortions. Other problems

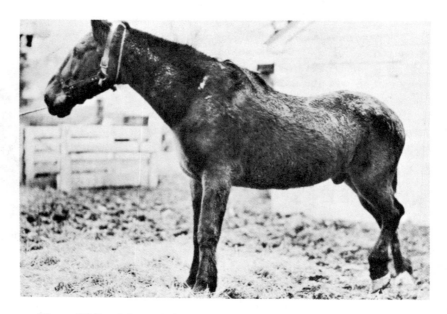

Figure 16-8   A horse infected with sleeping sickness. (United States Department of Agriculture)

associated with lymphoid tissues and the central nervous system have been observed. Treatment of the diarrhea and other symptoms is possible but no treatment is available against the virus itself. A modified live-virus vaccine is available.

Rickettsia

Rickettsia organisms were described by and named for Howard T. Ricketts in the early 1900s. With regard to size, rickettsia are intermediate to viruses and bacteria. They are similar to viruses in that they must live in other cells in order to survive. They are similar enough in other ways to bacteria that some scientists classify them as bacteria. Rickettsia are the causative organisms for some diseases in humans including Rocky Mountain spotted fever, Q fever, scrub typhus, and trench fever. Rickettsial diseases are usually transmitted by blood-sucking vectors such as ticks, lice, flies, and mosquitoes.

Anaplasmosis is a rickettsial disease of ruminants that is transmitted by a number of blood-sucking parasites. The disease attacks the red blood cells, thereby causing anemia. Anaplasmosis can also be transmitted from infected and carrier animals by unclean vaccinating, castrating, and dehorning equipment. The best preventive measures involve the control of ticks, flies, and mosquitoes. Treatment of the disease is difficult and expensive and is probably justified only for valuable breeding animals.

Protozoa

The term protozoa comes from Greek and means "first animal." These are one-celled organisms that were originally classified in the animal kingdom; however, in the newer classification systems, protozoa have been removed from the animal kingdom and placed in one of the other kingdoms.

Amebae are one kind of protozoa. They have no specific shape and move by extending a part of their cell membrane and causing the rest of the cell to flow into the extension. The extended part of the cell is called a pseudopod or false foot. This type of movement is called ameboid movement. White blood cells exhibit ameboid activity.

Some protozoa move about with the help of fine hairlike projections from the cell called cilia. The common paramecium found in pond water is a good example of such protozoa. The ciliates are not of major importance as causes of animal diseases.

The diseases trichomoniasis and African sleeping sickness are caused by protozoa that move by means of a relatively long whiplike projection called a flagellum. A kind of spore-forming protozoan known as coccidia form a protective microscopic cyst. They are eliminated from the body in the feces and eventually reinfect other animals. Coccidiosis is a protozoan disease that affects the inside lining of the digestive tracts of animals, particularly poultry.

Some protozoa reproduce by a process called budding. With this kind of reproduction, a part of the cell swells to form a bulge and ultimately breaks off to form a new cell. Other protozoa reproduce by a process similar to mitosis. A third kind of reproduction is by producing spores that develop into new cells. A classic example of a spore-producing protozoan is the one that causes malaria in humans.

Fungi

The fungi are a group of plantlike organisms that have no flowers, leaves, or chlorophyll. They include the mushrooms, yeasts, molds, and smuts. Many fungi are not microscopic; in fact, some mushrooms may grow to more than 2 ft in diameter. Fungi reproduce by producing spores either sexually or asexually. Sexual reproduction involves the production of male and female gametes. Some fungi, such as yeasts, may also reproduce by the method called budding.

Because fungi do not contain chlorophyll they cannot manufacture their own food. They must obtain their nutrients from living or dead plants and animals. Those that live off dead plant or animal matter are referred to as saprophytic fungi. Those that live off living material are called parasites. Many of the fungi that produce diseases in domestic animals and humans are parastic. Examples are ringworm, histoplasmosis, athlete's foot, and lumpy jaw.

Many fungi are beneficial. Some yeasts and mushrooms may be consumed as food. Most saprophytic fungi are beneficial in that they help in the decomposition of organic matter. Antibiotics such as penicillin and streptomycin are obtained from certain kinds of fungi.

WORMS AND ARTHROPODS

The study of worms, insects, arachnids, protozoa, and a few other organisms that infect animals is called parasitology. A parasite is an organism that lives in or on another animal called the host. The host does not benefit from the relationship but, rather, usually suffers from it. In the strict sense all microorganisms that cause disease could be called parasites, but bacteria and viruses are not usually included in a study of parasitology. Parasites may live inside the host, in which case they are called endoparasites or internal parasites. If they live on the outside of the body they are ectoparasites or external parasites.

Flatworms

Flatworms include the flukes and tapeworms and are classified under the phylum Platyhelminthes. The phylum name literally means "flat worm." Flatworms that infect domestic animals are internal parasites.

Liver flukes are flatworms that affect cattle, sheep, and goats. Eggs from flukes pass from their hosts in the feces. The eggs hatch in water and live in intermediate stages in snails. The snails are called intermediate or secondary hosts. Domestic animals are the primary hosts. Fluke larvae pass from the snails and become encysted on the stems and leaves of grass. Many of these cysts are eventually eaten by cattle and sheep. The flukes penetrate the intestinal walls and migrate to the liver. The length of their life cycle is about four months. Control measures involve interrupting the life cycle either in the primary or the intermediate host. Land drainage will help control the snails. Livestock can also be prevented access to the snail habitats. Treatment of fluke-infected animals should be carried out under the advice of a veterinarian.

Tapeworms live inside the intestines and absorb nutrients that have been digested by the host. The organism consists of a head part which is attached to the wall of the intestine and many flattened, blocklike segments that form a long chain or "tape." Tapeworms vary in size from those that are less than 1 in. long to those that may be 25 to 30 ft long. Each segment is a reproductive unit that can produce thousands of eggs. The eggs pass out of the host in the feces.

The eggs may hatch in water or inside other animals after they have been eaten. The newly hatched embryos are often called bladderworms. These embryos, or larvae, burrow through the intestinal walls and migrate to the muscle, lungs, brain, or other tissue. Bladderworms can cause severe damage to tissues and may even cause death.

Roundworms

Roundworms are sometimes called nematodes. Most of the roundworm parasites are classified under the phylum Nemathelminthes which means "round worm." Among the various parasites in farm animals, roundworms are among the most important. Several species exist, some of which may be only 1 or 2 mm in length while others may be 16 to 18 in. in length. Different species tend to inhabit different places in the body. Their particular location often leads to the names given them: stomachworms, bloodworms, kidneyworms, heartworms, and lungworms. Other worms in the nematode category are hookworms, ascarids, strongyles, and threadworms.

Because of the many kinds of worms that can affect farm animals and the many symptoms and problems that may result, it would be difficult to suggest any general treatment procedures to cover them all. It would be wise for owners of domestic animals to visit with their veterinarian to work out a particular worming program.

Insects

Insects are a large class of animals in the phylum Arthropoda. In fact, there are more species of insects in the world than all other animal species combined. Many insects are beneficial; only a relatively few species are harmful when we consider the vast number of species that exist. Insects have three pairs of jointed legs and three body regions. Many species of insects are winged in the adult stage. These insects are in a select class with the birds and bats as the only organisms that have the ability to fly.

Insect parasites include both endoparasites and ectoparasites. The internal parasites include the larval stages of a number of species of flies. Many of the ectoparasites are bloodsuckers such as mosquitoes, the biting flies, fleas, and sucking lice. In addition to causing problems for animals as parasites, many insects also act as vectors for a number of microbial diseases. A number of parasitic insects and some of their characteristics are discussed below.

1. Bloodsucking flies, also referred to as biting flies, include stable flies, horn flies, horse flies, and deer flies. These flies have sucking mouthparts equipped with special cutting parts to penetrate the tough skin of domestic animals. Stable flies are about the same size as house flies and to the inexperienced observer look very much like them. However, house flies do not bite. Horn flies are about one-half the size of stable flies. They get their name from their habit of resting on the base of the horns of cattle when not feeding. Horse flies are the largest of flies. Deer flies are smaller than horse flies but larger than stable flies.

Horse flies and deer flies lay their eggs near pools of water. The larvae live in or very near the water where they feed on other organisms in the mud. The larvae of stable and horn flies usually develop in manure piles. The life cycle of the horn fly may be as short as ten days. That of stable flies is about three weeks. Several broods of both species are produced each year. Horse flies have only one brood each year. Horse flies and stable flies can mechanically transmit anthrax, anaplasmosis, sleeping sickness, and certain other diseases.

Control of horse flies, deer flies, and stable flies is fairly difficult. Cattle can be sprayed with insecticide, but the flies remain on the animal for such a short period of time to feed that the chemical does not prevent them from biting and sucking blood. Horn flies are easier to control because of their habit of resting on the horns. Applications of insecticide around the neck and head every few weeks renders good control. The use of insecticides on livestock is regulated by the U.S. Food and Drug Administration which strictly specifies tissue residue limits of insecticides in livestock products. Label precautions should be observed when treating meat and dairy animals.

2. **Internal parasites** include screwworms, bots, and warbles, all of which are fly larvae that live parasitically in domestic animals. Screwworm flies lay their eggs on the edges of fresh wounds or moist places on the animals. After the eggs hatch, the larvae burrow into the live tissue and feed. They do not feed on dead or decaying tissue. During the fly season, livestock should be observed every few days to check for signs of infestation.

Screwworm flies have been almost completely eradicted from the United States. The eradication process involves the release of sterile male flies that mate with the females resulting in the production of infertile eggs that do not hatch. The females mate only once in their life so if mated to sterile males they will never lay fertile eggs. The flies are produced in large numbers in special laboratories where they are exposed to gamma radiation. This causes them to be sterile but otherwise they show no ill effects.

Cattle grubs are also called ox warbles. The adult is called the heel fly. The fly lays her eggs on the hair on the lower parts of the body. The eggs hatch and burrow through the skin. After several months of migrating through the body, they end up in the tissue under the skin along the back where they make a breathing hole through the skin. They spend one to two months here before emering and dropping to the ground to pupate in the soil. This pupa is similar to the cocoon in the life of a butterfly. After another one to three months the adult heel flies emerge to start the cycle over again. The adults do not bite or even feed, and live only about a week. Control measures depend in a large measure upon geographical location. Consult your veterinarian for local control measures. Many millions of dollars in damaged hides and contaminated meat are lost annually because of these flies.

Botflies lay their eggs around the head and neck of horses and sometimes other species. The eggs hatch when licked and burrow into the tissue inside the mouth. Within a few weeks they pass into the stomach. The bots spend about the next nine months attached to the walls of the stomach and intestines absorbing nutrients. The larvae finally pass out in the feces and pupate in the soil. The adults emerge, mate, and begin laying eggs. There is only one generation each year. Control measures involve treating the host to kill the bots inside.

3. Nonbiting flies include face flies and house flies. House flies do not normally affect farm animals although they may annoy them and carry filth. House flies lay their eggs in manure and almost any other kind of decaying waste or garbage. They are not guilty of being vectors of any specific disease, but could mechanically carry any disease organism. They are not bloodsuckers. If you have ever been bitten by what appeared to be a house fly, in all probability it was a stable fly.

Face flies are not bloodsuckers, but feed on the secretions around the eyes and nostrils. They are a common vector of the organisms that cause pinkeye. Face flies lay their eggs in fresh cow manure. Control of face flies is not as easy as that of horn flies. The faces of cattle must be treated every few days with a relatively safe or mild insecticide.

4. Fleas are generally not a problem to any domestic animals except dogs and cats. They are blood-feeding insects when in the adult stage. The larvae live in the gound or in the animal's bedding.

Commercial preparations of insecticides are available and are quite effective on dogs, but they should not be used on cats. If fleas are a problem on cats, a veterinarian should be consulted. Fleas are an occasional problem for swine and even humans. They are common on rodents. Fleas can act as vectors for a number of diseases of humans; one in particular is bubonic plague. Some recent outbreaks of the plague have occurred in the western part of the United States. Since the plaque-carrying flea is an ectoparasite of rats, plague control becomes essentially one of rat control. Ground squirrels and prairie dogs have also been known to carry these fleas.

5. **Lice** are wingless ectoparasites that usually spend their entire life on the body of the host (Fig. 16-9). Any species of domestic animal may become infected with lice. Some lice have chewing mouthparts. They feed on products from the skin including any available secretions from the body. Chewing lice are common on all species except swine. Sucking lice are equipped for piercing the skin and sucking blood. They are commonly found on most species of mammals. Sucking lice are not known to affect poultry. Swine are affected by only one species of louse, a sucking louse. Because of their habit of not leaving the host during their lifetime, lice are a bit easier to control than some other external parasites. Dipping or spraying with insecticides is effective. Devices such as backrubbers and dust bags may be used to help control lice, but they are not as effective as spraying or dipping (Fig. 16-10). Again, it is important to follow label directions and precautions strictly when treating farm animals.

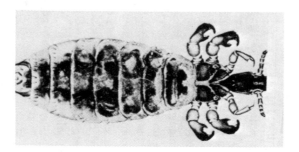

Figure 16-9 The long-nosed cattle louse. (Texas Agricultural Extension Service)

Figure 16-10   The most effective control for lice is dipping. (United States Department of Agriculture)

6. **Mosquitoes** are vectors of several diseases of livestock, especially of horses. Diseases carried by mosquitoes include equine encephalitis (sleeping sickness), equine infectious anemia (swamp fever), anaplasmosis, and fowl pox. Malaria in humans is transmitted by mosquitoes. Mosquitoes may attack livestock in such numbers as

to cause considerable blood loss as well as severe annoyance. The larvae of mosquitoes are strictly aquatic. Mosquito wigglers can be found anywhere there is standing water from tin cans and old tires to farm ponds and streams. One of the most effective ways to control mosquitoes is to eliminate the breeding sites or treat them with chemicals to kill the larvae.

Arachnids

Arachnids are another class of organisms included in the phylum Arthropoda. They have four pairs of jointed legs in the adult stage and are always wingless. The two most common parastic groups of arachnids are the ticks and mites (Fig. 16-11). Spiders are arachnids, but they are not usually a hazard to animal health. Except for two poisonous species, all spiders are generally beneficial. The two poisonous species are the black widow and the brown recluse or fiddleback spider.

Ticks cause irritation, discomfort, and blood loss. They are important vectors of a number of diseases including Rocky Mountain spotted fever, anaplasmosis, and tularemia (rabbit fever). Two families of ticks are the hard ticks and the soft ticks. The hard tick species are the most important as parasites of farm animals. Most of the common species are three-host ticks. The American dog tick is a

Figure 16-11   The sarcoptic mange mite. (Texas Agricultural Extension Service)

good example. Eggs are deposited on the ground where they hatch in two to three weeks into "seed ticks." These larval ticks may attach themselves to their first host and feed for several days. They then drop off their first host and molt to become nymphs.

The nymphs attach themselves to a second host and feed for about a week and drop to the ground again. After they molt this time they are adults. The adults crawl up on vegetation and wait for the third host. Mating usually takes place on the host. The females engorge themselves with blood and swell to a size that may be ten times their original size. After this engorgement, they drop to the ground to lay their eggs. Ticks in any of three stages may survive for several months without a blood meal. Tick control on livestock is very similar to that for lice.

Mites are quite small compared to ticks. They may infect all domestic animals. Some mites live on the skin and produce punctures that allows blood and tissue fluids to ooze out. The mites that cause mange and scabies burrow into the skin. The burrowing causes itching and inflammation of the skin. Serum and other fluids ooze from the skin to form scabs. Mites can cause considerable production losses and damage to hides. The chigger or red bug that attacks humans is the larval stage of a mite. It is also called the harvest mite and may affect birds and certain other animals. A number of systemic insecticides are available to control mange and scabies mites in livestock. A systemic insecticide is one that is absorbed by the body and circulates in the body fluids. A veterinarian should be consulted about their use.

NONINFECTIOUS DISEASES

Not all disease problems in domestic animals are due to the presence of organisms. According to our definition of disease, any defect in the body or body part qualifies as a disease. This would include such mechanical or physical problems as wounds and broken bones. Disorders due to deficiencies of vitamins, minerals, and hormones are other examples of noninfectious diseases. Also qualifying as diseases are problems caused by toxins from poisonous plants, and by pesticides.

Plant and Chemical Poisons

Poisonous plants have been known for centuries. References can be found in the Bible and in history books. Almost 2400 years ago, for example, Socrates drank a brew made from the hemlock plant after he had been sentenced to death by the judges of Athens. Some examples of plants that are toxic in varying degrees to cattle are the rayless goldenrod, jimson weed, larkspur, bracken fern, cocklebur seedlings, snakeroot, lupine, Japanese yew, and loco weed. Some of these same plants, as well as some others, are toxic to sheep. The list would include loco weed, lupine, rayless goldenrod, cocklebur seedlings, milkweed, greasewood, horsebrush, and rubber weed.

Common crop plants that are normally not poisonous may accumulate certain poisons. Two such examples of accumulated poisons are cyanide and nitrate. Cynaide (HCN) is also known as hydrocyanic acid or prussic acid. It may accumulate in certain sorghums such as Sudan grass. Potassium nitrate (KNO_3) can accumulate in certain grasses such as corn, oats, and sorghum when high amounts of nitrogen fertilizers are applied, especially in hot dry weather.

A number of chemicals can cause problems for livestock when consumed, even in relatively small amounts. Paint that contains lead, and lead from discarded storage batteries can cause problems. Because lead accumulates in the tissues, repeated consumption can eventually become lethal. Most of the pesticides can prove dangerous and even fatal to livestock. Compounds containing arsenic are sometimes used in weed killers, rodenticides, and insecticides. Insecticides and herbicides made from chlorinated hydrocarbons are dangerous because of their toxicity, but also because of their persistence. They have a high degree of resistance to biodegradability and are able to accumulate in the body. Some examples of chlorinated hydrocarbon insecticides are DDT, BHC, toxaphene, aldrin, dieldrin, chlordane, and heptachlor. Chlorinated herbicides include 2,4-D and 2,4,5-T.

Nutritional and Metabolic Diseases

A number of nutritional diseases are caused by deficiencies of vitamins and minerals. Some examples include rickets, osteomalacia, night blindness, and goiter. More complete lists of vitamin and mineral deficiencies are presented in Tables 11-3 and 11-4.

Bloat in cattle results from an accumulation of methane and carbon dioxide gases in the rumen. These gases, which are produced from the normal action of bacteria, are normally expelled from the rumen by belching. Bloat occurs when the gases cannot escape in this manner, or when they are produced in excess of what can normally be disposed of. Bloat causes the distension or swelling of the rumen which in turn causes obstruction of the large veins in the abdominal cavity or interferes with normal breathing. Death can occur within a few hours if the gases are not released. One way to release the gases is to pass a tube down the esophagus and into the rumen.

Even though the name of the disease suggests it, milk fever is not associated with an increase in body temperature. Milk fever can occur in high-producing cows at or shortly after parturition. It is associated with a severe hypocalcemia, the blood calcium dropping to one-third to one-half of normal levels. The affected animal will become partially paralyzed and lose consciousness. Within several hours a deep coma may develop leading to death. As soon as the symptoms of milk fever are suspected, a veterinarian should be called.

Grass tetany is a metabolic disturbance of cattle and sheep that most often occurs when animals are grazing lush pasture that is deficient in magnesium. Although it can occur in cattle and sheep of any age, it seems to occur more often in lactating cows turned into lush pasture after having been housed over the winter. The disease is associated with a low blood magnesium level. Nervousness, spasms, and convulsions are common symptoms. Death can occur within hours. A veterinarian should be called immediately upon noticing the first symptoms.

Genetic and Hormonal Diseases

Genetic diseases are quite often associated with deficiencies of certain enzymes or hormones, these deficiencies being caused by the absence of a certain gene. Phenylketonuria (PKU) in children is caused by the inability to properly metabolize the amino acid phenylalanine. The production of the enzyme that is necessary in this reaction is controlled by a dominant gene. If that gene is missing from the cells, PKU can result.

Another example of a genetic disease is a kind of dwarfism in cattle caused by the presence of two recessive genes. Because dwarf cattle are normally not used for breeding purposes, both the sire and the dam must carry the recessive gene in order for the dwarf to be produced. The parents are called carriers because they carry the recessive gene but do not express it. The genetic principles involved are explained in more detail in Chapter 5.

Several metabolic disturbances can result from either deficiencies or overproduction of certain hormones. For example, an overproduction of growth hormone can result in a giant. On the other hand, a deficiency of growth hormone can result in one type of dwarfism. A deficiency of insulin results in an increase in the blood sugar level called hyperglycemia. This condition can lead to the disease called diabetes mellitus or sugar diabetes. Other examples of disorders caused by hormone imbalance are presented in Chapter 3.

STUDY QUESTIONS AND EXERCISES

1. Define the following terms

active immunity	encephalomyelitis
acute	encephalitis
aerobic	epidemic
ameba	erysipelas
anaerobic	exudate
anemia	fever
anthrax	fission
antibiotic	flatworm
antibody	fluke
antigen	fungus
antiseptic	germ
antiserum	grass tetany
arachnid	health
aseptic	heel fly
bacillus	helminth
bacterium	host
blackleg	hydrophobia
bladderworm	immunity
brucellosis	immunology
budding	infection
chronic	infectious
cholera	infestation
coccidiosis	inflammation
coccus	insect
colostrum	larva
comfort zone	leptospirosis
contagious	louse
diarrhea	lymphocyte
disease	maggot
disinfectant	milk fever
ectoparasite	mite
edema	nematode

neutrophil
parasite
passive immunity
protozoan
pseudopod
pus
quarantine
rabies
rickettsia
roundworm
saprophyte
spirochete
spore

staphalococcus
streptococcus
tetanus
therapy
toxin
tuberculosis
undulant fever
vaccination
vaccine
vector
vibriosis
virus

2. Why is it important to have a basic understanding of animal diseases?

3. Distinguish between an acute and a chronic disease.

4. Give some examples of noninfectious diseases.

5. Explain the three links in the chain of disease transmission.

6. Compare disease transmission by direct and indirect modes.

7. Compare passive and biological transmission of disease.

8. What are some signs that might indicate than an animal is in poor health?

9. Compare the comfort zones of English and Indian breeds of cattle.

10. What is the body's first line of defense against infectious disease?

11. Describe the four general symptoms of the inflammation reaction.

12. Compare the functions of the various white blood cells.

13. Describe the antigen-antibody reaction.

14. Compare the various kinds of vaccines and how they are prepared.

15. Compare the characteristics of actively and passively acquired immunity.

16. Describe two examples of naturally acquired passive immunity.

17. List several ways in which bacteria may be beneficial.

18. Compare the cellular structure of a bacterium to that of a typical animal cell.

19. Indicate which of the bacerial diseases are caused by bacilli and which are not.

20. Which bacterial diseases affect the blood primarily?

21. Which bacterial diseases can lead to abortions?

22. Which bacterial disease is a waterborne disease?

23. What are some other names for brucellosis?

24. Name at least two anaerobic bacterial diseases.

25. Describe how a virus lives and reproduces.

26. How can viruses be controlled?

27. List several diseases that affect the nervous system.

28. How is rabies transmitted?

29. Which disease virus may be the smallest known virus?

30. What are some other names for equine encephalitis?

31. What do the letters BVD and IBR mean?

32. Compare the rickettsia to the viruses and bacteria.

33. Compare the various disease-producing organisms as to how they are classified biologically.

34. What is meant by ameboid activity?

35. Compare the several types of protozoa by their methods of movement.

36. Compare the various ways in which bacteria and protozoa reproduce.

37. Compare some of the characteristics of fungi to those of bacteria and protozoa.

38. Name some of the body tissues that may be attacked by roundworms.

39. Compare the body structure of insects and arachnids.

40. Compare the biting and nonbiting flies.

41. What precautions are necessary when treating livestock with insecticides?

42. Compare the life cycles of bot flies and heel flies.

43. Describe the irradiation method of screwworm fly control.

44. For what disease do face flies act as a vector?

45. Can you think of a reason why insecticides should not be used on cats?

46. For what disease are fleas a vector?

47. Compare the characteristics of ticks and lice.

48. Describe the two types of lice and the animals which they infest.

49. List some of the diseases for which mosquitoes are vectors.

50. What is the best way to control mosquitoes?

51. Describe the living and feeding habits of mites.

52. Describe the life cycle of the American dog tick.

53. What is a systemic insecticide?

Glossary

Abomasum: The fourth compartment of the stomach of a cow, sheep, or goat.

Absorption: The process whereby nutrients and other materials are taken through the walls of the digestive tract and into the blood.

Acrosome: The thin covering or cap over the head of a sperm cell.

Active immunity: The type of immunity in animals where the animal plays a part in the development of that immunity.

Acute: Usually refers to a disease that runs a short, severe course.

Adipose: Refers to the fat-filled cells of connective tissue.

Adrenalin: One of the hormones produced by the medulla of the adrenal glands; it is also called epinephrine. It helps in preparing the body for emergency actions.

Aerobic: Refers to organisms that require oxygen to live.

Afterbirth: The membranes of pregnancy which are expelled following parturition.

Air dry: Refers to feedstuffs that have been allowed to dry in the atmosphere; they usually contain between 5 and 15 percent water.

Agonistic: A type of animal behavior that involves offensive or defensive activities.

Albinism: A genetic condition in animals characterized by the complete absence of pigmentation in the skin, eyes, and hair.

Albumin: Also spelled albumen. The white of an egg; a type of protein.

Allantois: One of the three fetal membranes; located between the amnion and the chorion.

Alleles (allelomorphs): Genes that can occupy the same loci on a pair of homologous chromosomes, but have different effects.

Allergy: A rather severe reaction that results from the introduction of antigens into the body.

Alveolus: A microscopic saclike structure found in the lungs and the mammary glands.

Ameba: A protozoan organism with no specific shape that moves in a flowing manner by the extension of a false foot or pseudopod.

Amnion: The fetal membrane located nearest the fetus; it is filled with amnionic fluid to protect the fetus from shock.

Ampulla: A funnel-shaped structure located at or near the end of a duct; the part of the Fallopian tube between the infundibulum and the isthmus; a part of the vas deferens where it joins the urethra in some species.

Amylase: An enzyme secreted by the pancreas and delivered to the small intestine that aids in the digestion of starch.

Anaerobic: Usually refers to an organism that does not require oxygen to live and reproduce.

Anaphase: The stage of mitosis where the centromeres of the doubled chromosomes divide and separate producing two sets of chromosomes that move to the two poles of the cell prior to the division of the cytoplasm.

Anatomy: The branch of biology that deals with the structure of organisms.

Ancestor: An individual from whom an animal or person is descended.

Anemia: A deficiency of hemoglobin, iron, or red blood cells.

Anesthesia: Loss of sensation brought about by drugs, disease, or other factors.

Anestrus: The nonbreeding season; the period of time when a female is not cycling.

Anoxia: A deficiency of oxygen.

Anterior: In four-legged animals, toward the head; the opposite of posterior.

Anthrax: An infectious disease of domestic and other animals caused by spore-forming bacteria.

Antibiotic: A substance such as streptomycin or penicillin used to destroy or inhibit the growth of microorganisms.

Antibody: A substance synthesized by the body to counteract the effect of foreign substances (antigens) that enter the body.

Antidiuresis: A decrease in the production and excretion of urine.

Antigen: A foreign substance, usually protein, that when present in the blood causes the production of an antibody that is antagonistic to it.

Antiseptic: Refers to substances or procedures that prevent the growth and reproduction of microorganisms.

Antiserum: A serum that contains antibodies, which is used to treat or temporarily prevent certain infectious diseases.

Antrum: The cavity in a hollow organ or structure such as the cavity inside a developing follicle.

Appendage: An attachment to something larger; a leg or limb.

Arachnid: A group of arthropods having four pairs of legs and one or two body regions; mites, ticks, and spiders are examples.

Arteriosclerosis: A disease involving the thickening and hardening of the walls of the arteries.

Artery: A blood vessel that functions in carrying blood from the heart.

Artiodactyla: The zoological order of mammals that includes hoofed animals with an even number of toes such as cattle, sheep, and swine.

Aseptic: Free from living infectious or otherwise dangerous microorganisms.

Asexual: Without sex.

Assimilation: The process of converting food nutrients to body tissue.

Atheroma: A fatty degeneration of the inner walls of arteries.

Atherosclerosis: A disease involving the fatty degeneration of the inner walls of the arteries; a form of arteriosclerosis.

Atlas: The first cervical vertebra; it connects to the occipital bone of the cranium.

Atrium: One of the upper chambers of the heart.

Atrophy: A decrease in size of a tissue, organ, or other body part.

Autopsy: A post mortem examination involving the dissection of a body to learn the cause of death.

Autosomes: All the chromosomes except the sex chromosomes.

Avian: Refers to birds, the class Aves.

Axon: The long fiber of a nerve cell that carries nerve impulses away from the central nervous system.

Bacillus: A rod-shaped bacterium.

Backcross: A mating of an F_1 individual to one of the parental types or breeds.

Bacterium: One of a group of one-celled round, rod-shaped, spiral, or filamentous microscopic organisms.

Balanced ration: The daily food allowance for one animal prepared to include all of the required nutrients in the required proportions.

Barred: A trait in chickens involving the alternating white and dark stripes on the feathers.

Barren: Unable to produce offspring.

Barrow: A male pig castrated before reaching sexual maturity.

Basal feed: A feed used primarily for its energy content.

Bay: A color pattern in horses involving a brown or reddish brown body with black mane, tail, and legs from the hocks and knees down.

Bile: A substance produced by the liver and stored in the gall bladder; it functions in physically breaking down fat globules into smaller ones making it easier for them to be digested.

Biological value: A measure of the utilization of nitrogen by the body, expressed in percent; a measure of protein quality.

Bitch: A female dog.

Blackleg: An infectious disease of cattle caused by an anaerobic bacillus.

Bladderworm: The newly hatched embryos of certain tapeworms.

Bloat: A swelling of the rumen caused by an accumulation of gas.

Boar: A male pig.

Bolus: A soft, rounded mass of chewed food; a large pill.

Bomb calorimeter: An apparatus used to measure the amount of heat given off by any combustible substance.

Bovine: Refers to cattle.

Bran: A kind of feed made from the seed coat of wheat or other cereals.

Breed: A group of animals with certain common characteristics that distinguish them from others of the same species, and which they tend to transmit with reasonable consistency.

Breeding value: The ability of an animal to transmit the genetic capability to produce meat, milk, eggs, or other economically important products.

Breed true: To transmit a trait with reasonable consistency.

Brisket: The lower chest or breast of a quadruped.

Broad ligament: The connective tissue in the abdominal cavity that supports the reproductive tract in a mammal.

Broiler: Meat-type chickens about two or three months of age.

Brood: A group of baby chicks.

Broodiness: The nature or quality of a hen that causes her to want to set on a nest of eggs.

Brucellosis: A bacterial disease of domestic mammals caused by a bacillus which results in abortions; also called Bang's disease.

Buck: A male sheep, goat, deer, or rabbit.

Budding: An asexual form or reproduction in microorganisms where a part of the cell seems to be "pinched off" from the rest.

Bulky: Refers to a substance that has a relatively small weight per unit of volume.

Bull: A male of the cattle species.

Bullock: A steer.

Burro: A donkey.

Butter: A rather solid collection of fat globules obtained from churning cream and used for food.

Buttermilk: The milk that remains after butter has been obtained from the churning process; also made from skim milk by adding certain kinds of bacteria that produce the characteristic buttermilk flavors.

Caecum: *See* Cecum.

Calciferol: Another name for vitamin D_2 obtained from irradiated ergosterol.

Calf: A sexually immature member of the cattle species.

Calorie: A measure of the amount of heat required to raise the temperature of one gram of water from 14.5 to 15.5°C.

Candling: The process of examining the interior quality of an egg by holding it in front of a light; it was originally done using a candle.

Cannon: The part of the leg that contains the metacarpus or metatarsus.

Cannula: A tube inserted into a body or organ cavity to permit the escape of gas or fluids.

Canter: An easy-riding three-beat gait of a horse; a slow gallop.

Capon: A castrated male chicken.

Carbohydrate: A group of chemical substances made up of carbon, hydrogen, and oxygen, and taking the form of one or more sugar units.

Carcass: The body of a dead or slaughtered animal.

Cardiac: Refers to the heart.

Carnivorous: Refers to meat-eating animals such as dogs and cats.

Carrier: A genetic term that refers to an animal that expresses the dominant trait but is heterozygous for the recessive gene.

Cartilage: A kind of connective tissue found on the joint surfaces of long bones, between the vertebrae, and at other places in the skeleton.

Caruncles: Specialized points of attachment in the uteri of ruminants for fetal membranes; a fleshy outgrowth on the head and neck of a turkey.

Casein: The primary protein of milk.

Castrate: To remove the testes or ovaries.

Cattalo: A cross between the bison (buffalo) and domestic cattle.

Caudal: A directional term meaning toward the tail; the opposite of cranial.

Cecum: A pouch or blind gut attached to the side of the large intestine.

Cell: The building blocks of tissue containing a cell membrane, nucleus, and cytoplasm.

Cellulose: A carbohydrate that functions as a structural substance in the cell walls of plants; probably the most abundant substance in nature.

Centriole: A small organelle located near the nuclear membrane of cells that divides during mitotsis and forms the centers toward which the chromosomes move upon division of the cell.

Centromere: A small structure located on a chromosome that appears to form an attachment to the spindle fibers during cell division.

Cerebellum: A part of the brain located behind the cerebrum and above the brainstem that controls functions of intermediate complexity such as walking, running, and flying.

Cerebrum: The largest part of the brain; it controls body functions of the highest complexity.

Cervical: Refers to the vertebrae in the region of the neck.

Cervix: A part of the reproductive tract of female mammals that forms a seal or doorway between the uterus and the vagina.

Character: A distinguishing feature or quality; a synonym for trait.

Chestnut: A reddish or reddish brown color in horses with mane and tail of the same or lighter color; a horny or callous growth on the inner side of the legs of a horse.

Chick: A very young chicken.

Chigger: The larva of a mite that may burrow under the skin of humans and domestic animals and causes itching and inflammation.

Chlorophyll: The green material in plants that plays a role in the process of photosynthesis.

Cholesterol: A fat-soluble substance found in the fat, liver, nervous system, and other areas of the body; it plays an important role in the synthesis of bile, sex hormones, and vitamin D.

Chorion: The outermost of the three fetal membranes.

Chromatid: One strand of a doubled chromosome seen in the prophase and metaphase of mitosis; each strand is called a chromatid.

Chromosome: One of the several long thread-like structures found inside the nucleus of a cell; chromosomes are made of protein and DNA and carry the genes.

Chronic: Refers to a disease that is marked by long duration and frequent recurrence.

Cleavage: A kind of division in the developing embryo wherein the cells divide mitotically but their size is reduced with each division until they are the size of the usual body cell.

Clitoris: A small organ located at the ventral part of the vulva; it is homologous with the penis.

Cloaca: The common area or chamber that serves as the terminal part of the urinary, digestive, and reproductive tracts of birds, reptiles, and amphibians.

Close breeding: A form of inbreeding that involves very close relatives.

Clutch: The eggs laid by a hen on consecutive days.

Coccidiosis: A disease of the digestive tract caused by a group of protozoa called coccidia.

Coccus: A sphere-shaped bacterium.

Cock: A male chicken; also called a rooster.

Cockerel: A young male chicken.

Cod: That part of the scrotum that remains after castration.

Codominance: A kind of gene action where one allele does not exhibit complete dominance over the other.

Coefficient of digestibility: The amount of a particular nutrient that was digested and absorbed expressed as a percentage of the amount that was in the feed to begin with.

Coitus: Another word for copulation or sexual intercourse.

Colic: A digestive disturbance involving severe contractions, twisting, or obstructions in the digestive tract.

Collateral relative: Animals related not by being ancestors or descendants, but by having one or more common ancestors.

Colon: That part of the large intestine extending from the cecum to the rectum.

Colostrum: The first milk produced by a female mammal at the end of a pregnancy.

Colt: A young male horse.

Comb: The red fleshy growth on the top of the head of a chicken.

Comfort zone: The range of temperature within which an animal feels most comfortable and which makes no demand upon the animal's temperature-regulating mechanism.

Concentrate: A feed that is high in starch content and low in fiber; it may have high or low protein content.

Conception: The process whereby an animal becomes pregnant; fertilization.

Conformation: The shape and form of an animal.

Contagious: Refers to a disease that can be transmitted from one animal to another; infectious.

Copulation: Another word for coitus or sexual intercourse.

Corona radiata: The granular material adhering to an ovum after ovulation that gives it a sunburst appearance; cumulous oöphorus.

Corpus albicans: The small bit of scar tissue that remains on the surface of the ovary after the regression of the corpus luteum.

Corpus hemorrhagicum: The temporary blood clot that forms in the crater formed by the follicle after ovulation and prior to the development of the corpus luteum.

Corpus luteum: A structure that forms on the ovary following ovulation that functions in the secretion of progesterone.

Cortex: The outer layer or region of any organ.

Cortisol: A hormone of the adrenal cortex that functions in the metabolism of glucose and the reduction of certain kinds of stress.

Cotyledon: A structure on the fetal membranes that serves as a point of attachment to the maternal structure called the caruncle in the uterus of a ruminant animal; there may be as many as 100 such attachments on the fetal membranes.

Cow: A mature female of the cattle species.

Cranial: A directional term meaning toward the head; the opposite of caudal.

Cream: The layer of fat that rises to the top of unhomogenized milk.

Crest: The ridge on the neck of a mammal; the tuft of feathers on the heads of some birds.

Crossbreeding: The mating of animals of different breeds; breed-crossing.

Crossing-over: The exchange of parts between homologous chromosomes that occurs during the synapsis of the first division of meiosis.

Crude fiber: The part of feeds containing the cellulose, lignin, and other structural carbohydrates as determined by the proximate analysis.

Crude protein: A measure or estimate of the total protein in a feed determined by multiplying the nitrogen content by 6.25.

Cryotorchidism: A condition in mammals wherein the testes fail to descend into the scrotum; in horses, a ridgling.

Cud: A bolus of regurgitated food in ruminants.

Culling: The process of eliminating unwanted or poor-quality animals.

Curd: The coagulated part of milk after it sours or is treated with acid.

Cumulous oöphorus: The sticky granular material adhering to the surface of the ovum after ovulation; corona radiata.

Cytology: The science that deals with the origin, structure, and function of cells.

Cytoplasm: That part of the cell material located between the cell membrane and the nuclear membrane.

Dam: A female parent; mother.

Defecation: The process of eliminating feces from the body.

Dehorn: The surgical or chemical removal of the horns from cattle, sheep, or goats.

Deoxyribonucleic acid (DNA): The substance of which genes are made.

Dermatitis: An inflammation of the skin.

Dermis: The major layer of the skin which is located just under the epidermis.

Descendant: A living being that has obtained part of its genetic inheritance from another, known as its ancestor.

Deutoplasm: The nutritive material or yolk in the cytoplasm or vitellus of an ovum or egg.

Dewclaw: A vestigal toe that does not reach to the ground on the foot of a mammal.

Dewlap: The fold of skin that hangs from the throat and neck of some animals.

Dextrin: A kind of carbohydrate intermediate in size to the starches and sugars.

Diaphragm: The layer or sheet of muscle and connective tissue that forms the wall between the thoracic and abdominal cavities of mammals and aids in the process of breathing.

Diarrhea: An abnormal fluid defecation usually caused by an infection of the digestive tract.

Dicoumarol: An anticoagulant produced by molds on spoiled clovers.

Diestrus: The period of the estrous cycle that occurs between metestrus and proestrus.

Diet: The daily consumption of food or feed; ration.

Diethylstilbestrol (DES): A synthetic estrogenic hormone that has been used to stimulate faster growth and the deposition of additional fat in steers on feed.

Digestibility: The ability of food to be digested and absorbed by the animal rather than eliminated in the feces; expressed as a percentage of all food or food nutrients taken into the body.

Digestible energy (DE): The proportion of energy in a feed that can be digested and absorbed by an animal.

Digestible protein: The proportion of protein in a feed that can be digested and absorbed by an animal.

Digestion: The process whereby complex nutrients such as starch, fats, and protein are chemically broken down in the digestive tract to more simple nutrients such as glucose, fatty acids, and amino acids which can be absorbed and utilized by the body.

Dihybrid: An individual that is heterozygous for two pairs of genes.

Dihybrid cross: A cross or mating between two individuals that are heterozygous for two pairs of genes (*AaBb* × *AaBb*).

Diploid: Refers to cells that consist of pairs of homologous chromosomes.

Disaccharide: A kind of sugar made up of two monosaccharides bound together chemically.

Disease: The state of being where a part or parts of the animal body are not functioning normally.

Disinfect: To kill or otherwise render ineffective harmful micro-organisms and parasites.

Distal: Located in a position that is distant from the point of attachment of an organ; for example, the toes are located on the distal part of the leg.

Diuresis: An increased production of urine.

Dizygotic twins: Twins that develop from two separate fertilized ova.

DNA: *See* Deoxyribonucleic acid.

Dock: To remove the tail of sheep; the stub that remains after the tail of a sheep has been removed.

Doe: An adult female goat, rabbit, or deer.

Domesticate: To bring wild animals under the control of humans over a long period of time for the purpose of providing useful products and services; the process involves careful handling, breeding, and care.

Dorsal: Refers to the back, or toward the back; the opposite of ventral.

Draft animal: An animal used to pull or carry heavy loads.

Drake: A mature male duck.

Dressing percent: The carcass weight of a meat animal expressed as the percentage of the live weight.

Dry: The state wherein a mammal is not producing milk.

Duckling: A recently hatched duck.

Duodenum: The first or proximal region of the small intestine; that part connected to the stomach.

Ectoderm: The outer of the three basic layers of the embryo which gives rise to the skin, hair, and nervous system.

Ectoparasites: Parasites that spend all or a part of their life cycle on the external part of the body.

Edema: A condition where abnormally large amounts of water are present in the intercellular spaces of the tissues.

Efferent ducts: Very small ducts located in the testes that connect the rete testes to the epididymis.

Egg: The reproductive cell of a female; ovum.

Ejaculation: The discharge of semen from the reproductive tract of the male.

Embryo: A developing organism in the very early stages following the union of the ovum and sperm.

Embryology: The science that deals with the study of the embryo.

Embryo transplant: The removal of developing embryos from one female and their transfer to the uterus of another; it usually involves the superovulation of superior females and the eventual transfer of their embryos in an attempt to increase the number of superior offspring.

Encephalitis: An inflammation of the brain.

Encephalomyelitis: A disease involving an inflammation of the brain or spinal cord which is caused by a pathogenic organism; sleeping sickness.

Endocrine: Refers to glands without ducts, which release their secretions, called hormones, directly into the blood.

Endometrium: The inside lining of the uterus.

Endoparasite: A parasite that spends all or a part of its life cycle inside the animal body.

Entrails: The visceral organs of the body, particularly the intestine.

Environment: All of the nongenetic factors and conditions that affect the life and performance of an animal.

Enzyme: A large complex protein molecule produced by the body that stimulates or speeds up various chemical reactions without being used up itself; an organic catalyst.

Epididymis: A tube or duct connecting the testis to the vas deferens; it is located inside a sac or pouch along side the testis and functions in the storage and passage of sperm from the testis to the vas deferens.

Epinephrine: A hormone, also known as adrenalin, that is produced by the medulla of the adrenal glands.

Epiphysis: The head or cap or end of a long bone; it is separated from the diaphysis by the epiphyseal cartilage.

Epistasis: The type of gene action where genes at one locus effect or control the expression of genes at a different locus.

Epithelium: A kind of tissue that functions as a covering on other kinds of tissue such as the skin and the inner and outer coverings of organs.

Equine: Refers to horses.

Erection: The process whereby a body part is made to stand upright; a process where the penis becomes engorged with blood causing it to be firm and turgid.

Ergosterol: A cholesterol-like substance found in plants that, when irradiated with ultraviolet light, changes to vitamin D_2.

Erysipelas: A disease caused by a bacillus that usually affects the joints and causes difficult breathing; it primarily affects hogs.

Erythrocyte: A red blood cell.

Essential amino acid: An amino acid that cannot be synthesized in the body, or cannot be synthesized rapidly enough to meet the body's needs.

Estrogen: A hormone or group of hormones produced by the developing ovarian follicle; it stimulates female sex drive and controls the development of feminine characteristics.

Estrous cycle: The reproductive cycle in nonprimates; it is measured from the beginning of one estrus or heat period to the beginning of the next.

Estrus: The period during the estrous cycle when the female is capable of being fertilized and when she is sexually receptive to the male.

Ether extract: The fatty substances in food and feed that are soluble in ether; one of the procedures in the proximate analysis of feeds.

Ewe: A mature female sheep.

Exotic: Foreign, unfamiliar, new, imported; in animal agriculture, it refers to imported breeds, usually of cattle.

Exudate: Fluid that seeps from blood vessels into the intercellular space.

F_1: The first generation progeny of a P_1 cross; usually refers to an individual that is heterozygous for one or more pairs of genes.

F_2: The second generation progeny generally produced by crossing two F_1 individuals.

Fallopian tube: One of the two tubes or ducts connected to the uterus of mammals and leading to the ovary; functions in transporting the ovum from the ovary to the uterus.

Farrow: To give birth to a litter of pigs.

Feces: The material eliminated from the rectum of the digestive tract.

Feed: Materials given to animals to provide them with essential nutrients.

Feedstuff: Material used for feed.

Felial: Refers to the child or offspring; the meaning of the F in F_1 and F_2.

Femur: The large bone in the pelvic limb between the stifle and the pelvis.

Fertility: The ability to produce sperm, ova, or offspring.

Fertilization: The process wherein a sperm and an ovum fuse to form a diploid zygote.

Fetlock: The joint or region of the leg of a horse between the cannon and the pastern.

Fetus: A young unborn animal as it develops in the uterus of a mammal.

Fever: A body temperature above normal for the species.

Filly: A young immature female horse.

First meiotic division: The first of a series of two divisions in the process of producing haploid sex cells or gametes.

Fission: Reproduction by the separation of one cell or organism into two or more parts, each of which develops into complete organisms.

Fistula: An opening or passage way leading from the hollow part of an organ to the surface of the body; cannula.

Flatworm: One of the many organisms that are members of the phylum Platyhelminthes.

Fleece: The wool of a sheep.

Flesh: The skeletal muscles of an animal.

Flock: A group of sheep or birds.

Fluke: A flattened worm that is often found as an animal parasite.

Foal: A young, unweaned horse or mule of either sex.

Fodder: A coarse feed, such as corn stalks, fed to cattle and horses.

Follicle: A small anatomical cavity; particularly, a small blisterlike development on the surface of the ovary that contains the developing ovum.

Follicle-stimulating hormone (FSH): A hormone produced by the anterior lobe of the pituitary gland that stimulates the growth and development of the follicle of an ovary, or the development of sperm in the testes.

Forage: A kind of roughage such as grasses and legumes used for animal feed; it may be fed as pasture, hay, or silage.

Foramen: A small opening, usually in bone.

Founder: An inflammation of the tissue that attaches the hoof to the foot; it may be caused by overfeeding, concussion, or a number of other factors.

Fowl: Refers to a bird, usually poultry.

Foxtrot: An uneven, easy-to-ride, four-beat gait in horses, intermediate to the walk and trot.

Freemartin: A sterile heifer born twin to a bull calf.

FSH: *See* Follicle-stimulating hormone.

Fungi: Plant-like organisms that have no chlorophyll; they get their nourishment from living or decaying organic matter.

Gait: The action of a horse's legs, such as walking or running.

Galactophore: A milk duct of the mammary gland.

Gamete: A reproductive or sex cell that contains one-half of the genes and chromosome of the parent; a sperm or an ovum.

Gametogenesis: The process in plants or animals, male or female, involving the production of gametes; ovigenesis or spermatogenesis.

Gander: A mature male goose.

Gastric: Refers to the stomach.

Gelding: A castrated male horse.

Gene: The smallest unit of inheritance found as a part of a chromosome.

Generation interval: The period of time between the birth of one generation and the birth of the next.

Genetic correlation: A situation where two or more traits are affected by many of the same genes.

Genetics: The science that deals with heredity and variation in organisms, and with the function and transmission of genes.

Genital: Refers to the reproductive organs.

Genotype: The genetic makeup of an animal.

Germ: A general term for any disease organism; an embryonic or seedlike organism.

Gestation: The period in the life of a mammal from fertilization to parturition; pregnancy.

Gilt: An immature female hog.

Gland: A small organ that secretes a useful fluid.

Globule: A collection of several molecules of fat that takes on a spherelike appearance, and is insoluble in water.

Glucagon: A hormone produced by the pancreas that stimulates a rise in blood sugar.

Glucose: A common monosaccharide sugar that serves as the building block for many larger carbohydrates; blood sugar.

Gluten: A kind of protein found in wheat and other cereal grains.

Glycerol: One of the components of a fat molecule; a fat molecule is composed of three fatty acids attached chemically to glycerol.

Goiter: An enlargement of the thyroid gland usually brought on by a deficiency of iodine.

Gonad: The organ in a male or female animal that produces the gametes; an ovary or a testis.

Gonadotropin: A hormone that stimulates the gonads, particularly, FSH and LH.

Gosling: A very young or recently hatched goose.

Grading up: The process of improving grade or mediocre animals by the continuous use of purebred sires.

Grass: A kind of plant with long slender, bladelike leaves that produces monocotyledonous seeds; examples are corn, wheat, and fescue.

Grass tetany: A disease of cattle caused by magnesium deficiency, and resulting in nervousness, muscle spasms, and convulsions.

Graze: To consume growing forage.

Grease wool: Raw wool after it is removed from a sheep and before it is scoured.

Green chop: Forage that has been freshly cut and immediately fed to livestock.

Gregarious: Having the tendency to associate with others of like kind, as the tendency of sheep to flock together.

Gristle: Cartilage.

Gross energy (GE): The total amount of heat or energy in an organic substance as measured by complete combustion in a bomb calorimeter.

Habituation: The process of becoming accustomed to a particular environment.

Half-sib: A half brother or sister.

Hand: A unit of measure used to express the height of horses from the withers to the ground; it is the approximate width of a man's hand, four inches.

Haploid: Refers to a cell that has one chromosome from each pair of homologous chromosomes that exist in a diploid cell.

Hatch: The act of the young coming forth from an egg as with birds; a process that is somewhat analogous to parturition in mammals.

Hay: Livestock feed made from forage that has been cut and allowed to dry so that it may be stored with a minimum of deterioration.

Haylage: Forage that could have been cut for hay but is put up with a higher moisture content than hay, but with less moisture than silage.

HCG: *See* Human chorionic gonadotropin.

Health: The state wherein all body parts are functioning normally.

Heart girth: An imaginary line around the body of an animal just behind the shoulders; a measurement that is sometimes used to estimate body weight.

Heat increment: The amount of heat or energy produced by an animal in eating, digesting, absorbing, and metabolizing food; also called the work of digestion.

Heat period: Estrus; the period during which a female is sexually receptive.

Heel fly: The adult form of the cattle grub or ox warble.

Heifer: The young female of the cattle species; usually applies to the female that has not yet had a calf.

Helminth: A worm or wormlike parasite.

Hemocytometer: A specially designed microscope slide used to count red blood cells.

Hemoglobin: The red substance in the red blood cells that functions in carrying oxygen from the lungs to the tissues.

Hemolysis: A process that results in the rupture and release of hemoglobin from red blood cells.

Hemophilia: A genetic disease wherein the blood does not clot properly.

Hen: An adult female chicken or turkey.

Herbivore: An animal that feeds primarily on plant materials.

Herd: A group of animals, particularly cattle, horses, or hogs.

Heredity: The transmission of genetic material from one generation of animals to the next.

Heritability: That proportion of the total variation for any trait that is due to genetic causes, and that can be expected to be passed on to the next generation.

Hermaphrodite: An animal that possesses complete sets of male and female reproductive organs and can produce both male and female gametes; true hermaphrodites do not exist in birds and mammals.

Hernia: The protrusion of internal organs through an opening in the body wall, as with a scrotal or umbilical hernia.

Heterogametic: Refers to the sex (male or female) that possesses two kinds of sex chromosomes; in mammals, the heterogametic sex is the male because it has both an X and a Y chromosome.

Heterosis: The amount of superiority observed or measured in crossbred animals compared to the average of their purebred parents; hybrid vigor.

Heterozygote: An individual whose genotype is made up of two different genes (alleles) occupying the same loci on a pair of homologous chromosomes; the individual may be heterozygous at one or more loci.

Hilus: The stalk of the ovary that serves as the attachment to the broad ligament.

Hinny: The offspring of a female donkey bred to a stallion.

Histoplasmosis: A disease of the lymph nodes in the region of the neck caused by a fungus.

Hock: The joint on the pelvic limb of a horse between the gaskin and the cannon; same as the joint in the pelvic limbs of other quadrupeds.

Homeostasis: The combination of body mechanisms that causes the body to maintain a constant internal environment.

Homeothermic: Refers to animals that are able to maintain a fairly constant body temperature; warm-blooded.

Homogametic: Refers to the particular sex of the species that possesses two of the same kind of sex chromosome such that only one kind of gamete can be produced with respect to the kinds of sex chromosomes it contains; in mammals, the female is the homogametic sex (XX).

Homogenized milk: Milk that has been treated in such a way as to break up the particles or globules of fat to a size small enough that they will remain suspended and not rise to the top after standing.

Homolog: One of a pair of structures having similar structure, shape, and function, as with two homologous chromosomes.

Homologous chromosomes: Pairs of chromosomes that are the same length, that have their centrioles in the same position, and that pair up during synapsis in meiosis.

Homozygote: An individual having identical genes at homologous loci, as with *AA* or *aa;* the individual may be homozygous at several loci.

Hook bones: The prominent bones on the back of a cow formed by the anterior ends of the ilii (plural of ilium); points of the hip.

Hormone: The general name for the secretion of an endocrine gland.

Host: An animal upon which another organism lives as a parasite.

Human chorionic Gonadotropin (HCG): An LH-like hormone secreted by the human placenta.

Humerus: The large bone of the pectoral limb located between the elbow and shoulder joints.

Husbandry: The art and science of cultivation and production of plants and animals.

Hybrid: A heterozygote; the offspring of a cross between two genetically unrelated parents.

Hybrid vigor: Same as heterosis.

Hydrophobia: Rabies; a viral disease that affects the brain and results in an inability to swallow and, thus, an inability to drink.

Hyperglycemia: An above-normal blood sugar level.

Hypocalcemia: A below-normal level of calcium in the blood.

Hypodermis: The layer of tissue below the dermis of the skin.

Hypophysectomy: The surgical removal of the pituitary gland.

Hypophysis: Another term for the pituitary gland, a structure located on the underside of the brain and embedded in the sphenoid bone.

Hypostatic: The trait whose expression is modified in the case of epistatic gene action.

Hypothalmus: The part of the brain attached to the brainstem; the central part of the brain through which the nerve fibers pass from the brainstem on their way to the cerebrum.

Hysterectomy: The surgical removal of the uterus.

Ice cream: A frozen dairy food containing approximately 10 percent fat and 15 percent sugar, flavoring, and usually eggs.

Ice milk: A frozen dairy product similar to ice cream except that it does not contain as much fat.

Identical twins: Twins that develop from a single fertilized egg that separates into two parts shortly after fertilization.

Ileum: The last division of the small intestine that joins the large intestine.

Immunity: The ability of an animal to resist or overcome a particular disease.

Implantation: The process whereby the mammalian embryo forms an attachment to the uterine wall.

Impregnate: To fertilize a female animal; to make pregnant.

Imprinting: A kind of behavior common to some newly hatched birds that causes them to adopt the first animal or object they see as their parent.

Inbreeding: A type of mating system where the animals that are mated are more closely related than the average of the population.

Incomplete dominance: A kind of inheritance where a gene does not completely cover up or modify the expression of its allele; also may be known as codominance or blending inheritance.

Infection: A condition wherein an animal becomes infested with disease organisms.

Infestation: A condition wherein an animal is invaded by insects, mites or ticks; similar to infection.

Inflammation: The reaction of tissue to an injury to limit or control it; it is characterized by swelling, redness, fever, and pain.

Infundibulum: The enlarged, funnel-shaped structure on the end of the Fallopian tubes that functions in collecting the ova during ovulation.

Ingestion: The process of taking food into the mouth; eating.

Inguinal canal: The opening in the abdominal wall through which the testes pass from the body cavity into the scrotum.

Insect: A class of arthropods that have six legs and three definite body regions—the head, thorax, and abdomen.

Instinct: The ability of an animal based upon its genetic makeup to respond to an environmental stimulus; it does not involve a mental decision.

Insulin: A hormone produced by the pancreas that stimulates cells to metabolize blood sugar and usually results in a reduction of blood sugar levels.

Interphase: The period in the life of a cell between mitotic divisions.

Intestine: The part of the digestive tract between the stomach and anus.

Involution: The return of an organ to normal size after a time of enlargement, as in the case of the uterus after parturition.

Islets of Langerhans: Specialized tissue in the pancreas that secretes insulin.

Isthmus: In biology, a narrow piece of tissue that connects two larger parts; the part of the Fallopian tube between the ampulla and the uterus.

Jack: A male donkey.

Jenny: A female donkey.

Jejunum: The part of the small intestine between the duodenum and the ileum.

Jowl: The cheek of a pig.

Karyotype: A picture or diagram of the chromosomes of a particular cell as they appear in the metaphase of mitosis arranged in pairs by size and location in the centromeres.

Keratin: A kind of protein found in hair, feathers, horns, and hoofs.

Ketosis: A metabolic disease characterized by excessive amounts of ketone bodies such as acetone in the blood and body fluids.

Kid: An immature goat.

Kjeldahl: Refers to a procedure for determining the amount of nitrogen in organic materials; named for the Danish chemist who developed it.

Lactation: The process of producing or secreting milk.

Lamb: An immature sheep; to give birth to a lamb.

Lanolin: An oily substance obtained from raw wool that is used in cosmetics and ointments.

Lard: Fat obtained from pork.

Larva: The immature form of an insect, worm, or other lower forms of animals; it must undergo a considerable change in form before becoming adult.

Larynx: The part of the trachea nearest the pharynx.

Lateral: A directional or positional term meaning away from the middle or toward the side.

Legume: A plant that contains nodules on the roots that contain special kinds of bacteria that are able to convert atmospheric and soil nitrogen into protein; peas, beans, clover, and alfalfa are legumes.

Leptospirosis: A bacterial disease of the kidney caused by a spirochete.

Lesion: An abnormality of a tissue or organ caused by injury or disease.

Lethal gene: A gene that can cause the death of an individual when it is allowed to express itself.

Leukocytes: White blood cells.

Leydig cells: Special cells or tissue located inside the testes that secrete testosterone; also called interstitial cells.

LH: *See* Luteinizing hormone.

Libido: Sex drive.

Lignin: A complex, highly indigestible material associated with cellulose and the other fibrous parts of a plant.

Linebreeding: A kind of inbreeding that attempts to concentrate the genes of a particular ancestor into the pedigree.

Linecross: A cross between two inbred lines.

Linkage: Refers to two or more genes located on the same chromosome.

Lipase: A fat-digesting enzyme.

Lipid: A group of organic substances that are insoluble in water but soluble in such materials as acetone, chloroform, ether, and xylene; examples include fats, waxes, and steroids including cholesterol.

Litter: A group of animals born at the same time from a single mother; litter-bearing animals include swine, dogs, cats, rabbits, and rats.

Livestock: Farm animals raised to produce milk, meat, work, and wool; includes beef and dairy cattle, swine, sheep, horses, and goats; may also include poultry.

Locus: The position or region of a chromosome where a gene is located.

Loin: The region or part of the back between the thorax and the pelvis.

Louse: A small wingless insect that usually lives as a parasite on humans or domestic animals.

Lumbar: The region of the back between the thorax and the pelvis; refers to the loins.

Lumen: The cavity inside a tubular organ.

Luteinizing hormone (LH):   A hormone of the anterior pituitary gland that stimulates ovulation and development of the corpus luteum in females and secretion of testosterone by the interstitial cells in males.

Lymph: The fluid that flows through the lymphatic vessels.

Lymphocyte: A kind of white blood cell produced by lymph nodes and certain other tissues, and associated with the production of antibodies.

Lysozyme: An enzyme present in certain body secretions that can destroy certain kinds of bacteria.

Maggot: The larva of a fly.

Magnum: The part of the oviduct of a bird located between the infundibulum and the isthmus; the source of the albumin of an egg.

Mammal: A group of animals that secrete milk, and nurse their young, grow hair on their bodies, and possess a diaphragm between the thoracic and abdominal cavities.

Mandible: The large, lower bone of the jaw.

Marbling: The interspersion of fat among the muscle fibers of meat.

Mare: A mature female horse.

Mastication: The process of chewing the food.

Mastitis: An inflammation of the mammary glands caused by microorganisms or mechanical injury.

Maxilla: The smaller, upper bone of the jaw that contains the upper molars.

Meadow: Land upon which grasses and legumes are grown for hay or pasture.

Mean: A value determined by the addition of several values and dividing by the number of values added; the average.

Meat: Food composed of the skeletal muscles of animals.

Medial: A directional term that means toward the middle.

Median: The middle.

Median plane: A plane that runs from the anterior to posterior ends of an animal that separates it into two equal parts.

Medulla: The inner layer or part of an organ.

Meiosis: A type of cell division associated with the production of sex cells or gametes; the process begins with a diploid cell and results in the production of haploid cells.

Melanin: The black or dark brown pigment found in skin and hair cells.

Meninges: The membranes that cover the central nervous system.

Menstrual cycle: The reproductive cycle in primates as measured from the beginning of one menstruation to the beginning of the next, provided pregnancy has not occurred.

Menstruation: The discharge of blood and tissue that developed in the uterus in preparation for pregnancy in primates; this process occurs during each cycle unless pregnancy has occurred.

Metabolizable energy (ME):   The amount of energy in food after subtracting or accounting for that in the feces, urine, and, in the case of ruminants, the energy in the rumen gases.

Metaphase: The phase of cell division between prophase and anaphase wherein the chromosomes become aligned on the equatorial plane of the cell.

Metestrus: The phase of the estrous cycle of nonprimates following estrus and characterized by the development of the corpus luteum and the preparation of the uterus for pregnancy.

Milk fever: A disease of cows that occurs at about the time of parturition which involves a slight generalized paralysis of the muscles; it is associated with a low level of calcium in the blood.

Mite: A group of arthropods that live as parasites on animals and cause mange and scabbies.

Mitosis: A kind of cell division wherein a diploid cell divides to produce two identical diploid cells; it is associated with the growth and maintenance of body tissue.

Monoestrous: Refers to an animal that has only one estrous cycle each year.

Monogastric: Refers to an animal that has only one stomach or stomach compartment.

Monohybrid: An individual that is heterozygous for one pair of genes.

Monohybrid cross: A cross between two individuals that are heterozygous for one pair of genes; an example is $Aa \times Aa$.

Monorchid: A male with one testis in the scrotum and one inside the abdominal cavity.

Monosaccharide: The simplist form of sugar; a single sugar unit.

Mortality: Death or death rate.

Mule: The animal that results from a cross between a mare horse and a jack.

Muley: Polled.

Multiple alleles: More than two different genes that can occupy the same locus on homologous chromosomes.

Mutation: A change in a gene resulting in the formation of an allele.

Mutton: The meat obtained from a mature sheep.

Muzzle: The nose and mouth of an animal.

Myometrium: The muscular layer of the uterus which brings about the expulsion of the fetus at parturition.

Nematode: A roundworm.

Net energy: The amount of energy in a feed after essentially all losses have been accounted for, particularly the losses in the feces, urine, digestive gases, and the heat increment; the difference between the metabolizable energy and the heat increment.

Neuron: A nerve cell.

Neutrophil: A phagocytic white blood cell associated with the formation of pus.

Nitrogen-free extract (NFE):   The portion of a feed made up primarily of starches and sugars; in the proximate analysis, NFE is determined by subtracting the ether extract, crude fiber, crude protein, ash, and water from the total weight of the sample.

Nonadditive genes: Genes that express themselves in a dominant or epistatic fashion.

NPN: Nonprotein nitrogen, such as urea.

Nucleus: That part of a cell containing the chromosomes and enclosed by a special membrane.

Nutrient: A substance or element found in feeds that is needed in the metabolic processes of the body.

Nutrition: The process of providing, digesting, and metabolizing nutrients.

Oil meal: A kind of feed obtained from soybeans, cottonseed, flaxseed, and certain other seeds after the oil has been removed, and which contains a relatively large amount of protein.

Omasum: One of the smaller compartments of the ruminant stomach located between the rumen and the abomasum.

Omnivore: An animal that eats food of both plant and animal sources.

Oöcyte: Ovicyte; one of the intermediate cells in the process of ovigenesis.

Oögenesis: Ovigenesis; the process of producing the female gamete, the ovum.

Oögonium: Ovigonium; the first or primary germ cell from which the female gamete is produced.

Open: Not pregnant.

Osmosis: The process whereby a fluid will pass through a semi-permeable membrane from a lower to higher concentration of solution in an attempt to equalize the concentrations.

Osteocyte: A bone cell, particularly one encased in hard bone.

Osteomalacia: A disease characterized by a weakening of the bones due to a deficiency of vitamin D or of calcium and phosphorus; usually occurs in mature animals and is similar to a disease in young animals called ricketts.

Outbreeding: A mating system involving individuals that are less closely related than the average of the population; the most common example is the mating of animals from two different breeds, called crossbreeding.

Outcrossing: A form of outbreeding involving relatively unrelated animals from the same breed.

Ovary: The primary organ of reproduction in the female; it produces the female gamete (ovum) and the female sex hormones.

Overdominance: A kind of interaction between alleles in which the heterozygote is superior to either of the homozygotes.

Ovicyte: Same as oöcyte, an intermediate cell in the process of ovigenesis.

Oviduct: The reproductive tract in birds; another name for a Fallopian tube.

Ovigenesis: Oögenesis; the process of producing the female gamete.

Ovigonium: The primary germ cell from which the female gamete is produced.

Ovine: Referring to sheep.

Oviparous: The production of offspring from eggs that hatch outside the body of the dam.

Oviposition: The process of laying an egg.

Ovoviparous: Refers to animals who produce eggs that are incubated inside the body of the dam and hatch inside the body or shortly after laying.

Ovulation: The process of releasing eggs or ova from the ovarian follicles.

Ovum: The female gamete or reproductive cell.

Ox: A member of the bovine family; more often refers to a steer used for draft purposes.

Oxytocin: A hormone of the posterior pituitary gland that functions in the release of milk from the mammary gland and aids in parturition.

P_1 : A parental cross, usually involving individuals homozygous for different alleles, such as $AA \times aa$ or $AABB \times aabb$.

Paint: A color pattern in horses involving white patches on a dark background.

Palatability: The degree to which a feed is liked or acceptable to an animal.

Palomino: A copper, blond, or golden color in horses, usually with a lighter mane and tail.

Pancreas: An organ located near the duodenum of the small intestine that secretes digestive enzymes and the hormones insulin and glucagon.

Parasite: An organism that spends at least a part of its life cycle on or in the body of another animal called the host.

Parathormone (PTH): The hormone of the parathyroid glands; helps to maintain the calcium level of the blood by removing calcium from the bones; also called parathyroid hormone.

Parental generation: The P_1 generation; the first generation in a series of crosses; usually involves homozygotes for different alleles.

Parthenogenesis: A kind of reproduction where an egg develops into an organism without having been fertilized.

Partial dominance: A kind of interaction between alleles where one gene is not completely dominant to its allele but where the appearance of the heterozygote is more similar to one of the homozygotes than to the other.

Parturition: The act of giving birth in mammals.

Passive immunity: A kind of immunity acquired by animals when they are injected with antibodies against some disease; the animal plays no part in the development of the immunity.

Pastern: The region of the foot or leg between the hoof and the fetlock joint.

Pasteurization: A process where milk is heated to 72°C for 15 seconds or to 63°C for 30 minutes to kill dangerous organisms, without causing the milk to curdle; named for Louis Pasteur.

Pasture: Grasses and legumes grown for grazing animals.

Pathogen: A disease-producing organism.

Pathology: The science that deals with diseases and the effects that diseases have on the structure and function of tissues.

Paunch: Another name for the rumen.

Pedigree: A list or chart showing the names of an animal's ancestors.

Penetrance: The proportion of time that a phenotype is expressed compared to how often it is expected.

Penis: The copulatory organ of the male.

Pepsin: Gastric protease; the protein-digesting enzyme of the simple stomach and abomasum.

Peptide: A compound made up of a series of amino acids; an intermediate in the synthesis or digestion of a protein.

Pericardium: The sac or membrane that encloses the heart.

Periosteum: The outer membrane or covering of bone.

Peristalsis: Specialized contractions of a tubular organ that causes materials inside the tube to move along.

Peritoneum: The inner lining of the abdominal cavity.

pH: A symbol that refers to the degree of acidity or alkalinity of a fluid; defined chemically as the negative logarithm of the hydrogen ion concentration; the pH scale ranges from 0 to 14 with 7 being neutral.

Phagocytosis: The process whereby certain cells such as some white blood cells and amebae engulf microorganisms and other particles.

Pharynx: The cavity that connects the mouth and nasal cavity to the throat; a passage common to the digestive and respiratory tracts.

Phenotype: The appearance of an animal or one of its traits; the way an animal looks or behaves as determined in part by the genotype.

Photosynthesis: A process in plants whereby the energy of the sun is used to convert water and carbon dioxide into sugars.

Physiology: The science that deals with the function of the body and its organs, systems, tissues, and cells.

Piebald: A color pattern in horses involving a black coat with white spots.

Pinbones: In cattle, the posterior ends of the ischii that appear as the two raised areas on either side of the tail head.

Pinocytosis: The engulfing or absorption of fluids by cells.

Pinto: A color pattern in horses involving white spots on any color background.

Placenta: The regions or area of attachment between the fetal membranes and the endometrium of the uterus during pregnancy; carbon dioxide and other wastes are absorbed by the dam through the placenta, and oxygen and other nutrients are absorbed from the dam through the placenta by the fetus.

Placentome: One of many structures in the pregnant uterus of a ruminant that make up the placenta; composed of a caruncle of the uterus and a cotyledon of the fetal membranes.

Pleiotropy: A situation where one gene affects more than one trait.

PMS: Pregnant mares' serum; usually contains a hormone with activity very much like that of FSH.

Poikilothermic: Refers to animals such as reptiles, amphibians, fish, and insects that are not able to maintain a constant body temperature; the body temperature normally varies with that of the environment.

Poll: The region at the crest or apex of the skull in horses and cattle.

Polyestrous: Refers to an animal that has several estrous cycles in a breeding season.

Polygastric: Having many stomach compartments as in the ruminant animal.

Polyneuritis: Inflammation of many nerves.

Polyploidy: Having more than two sets of homologous chromosomes; the symbols 3n, 4n, etc., are used to refer to triploidy, tetraploidy, etc.

Polysaccharide: A large molecular weight carbohydrate made up of many sugar units; examples are starches, cellulose, and glycogen.

Polyspermy: The entrance of many sperm cells into the ovum at fertilization.

Pony: A small horse, usually shorter than 14½ hands (58 in.) at maturity.

Porcine: Refers to swine.

Posterior: Refers to the back or caudal part; opposite of anterior.

Postnatal: Following birth.

Postpartum: Following parturition.

Poult: An immature turkey.

Poultry: Birds raised for meat and eggs.

Precurser: A compound from which another is made or synthesized.

Predator: An animal that preys upon and devours other animals.

Pregnancy: The period during which an embryo or fetus is developing inside the uterus; measured from fertilization to parturition.

Prehension: The taking of food into the mouth.

Prepotent: The ability of an animal to transmit its characteristics to its offspring with a high degree of success.

Primate: A member of a group of mammals including humans, monkeys, and apes.

Proestrus: The phase of the estrous cycle just before estrus; characterized by the development of the ovarian follicle.

Progeny: The offspring of an animal.

Progeny testing: Determining the breeding value of an animal by studying its progeny.

Progesterone: A hormone produced by the corpus luteum of the ovary that functions in preparing the uterus for pregnancy and maintaining it if it occurs.

Prolactin: A hormone of the anterior pituitary gland that functions in stimulating the secretion of milk.

Prolapse: Abnormal protrusion of the inner part of any organ.

Prolific: Having the ability to produce many offspring.

Prophase: The first phase of cell division wherein many of the preparatory steps take place such as shortening and thickening of the chromosomes, division of the centromeres, disappearance of the nuclear membrane, and formation of the spindle.

Prophylaxis: The prevention of disease.

Prostaglandins: A group of hormonelike substances produced by the vesicular glands and other tissues, having a wide range of biological effects.

Prostate: One of the accessory glands of the male reproductive system that encircles the neck of the bladder where it joins the urethra.

Protein: A complex substance made up of amino acids; found in the structural parts of animal cells as well as in parts of the cytoplasm; constitutes approximately two-thirds of the cell solids.

Protein supplement: Animal feed that contains approximately 20 percent or more protein.

Protozoa: A group of one-celled organisms that generally do not contain chlorophyll, including amebae, paramecia, flagellates, and certain spore-forming organisms; sometimes classified as one-celled animals.

Proximal: Opposite of distal; near the point of attachment or reference.

Proximate analysis: A system of analysis used to determine the total composition of nutrients in feed.

PTH: *See* Parathormone.

Puberty: The age at which an animal becomes sexually mature, producing viable gametes and sex hormones.

Pullet: A young female chicken.

Purebred: An animal that is produced as a result of several generations of breeding among animals of a given breed or line; they usually tend to be more homozygous than animals that are not purebred.

Purebreeding: The practice of breeding animals from within the same breed or line; the production of purebreds.

Pure strain: An animal that is similar to a purebred but the breeding program usually involves a greater degree of inbreeding.

Pus: The material produced at the site of an infection consisting of tissue fluids, white blood cells, dead tissue cells, and microorganisms.

Quadruped: An animal that walks on all four appendages.

Qualitative traits: Traits having a sharp distinction between phenotypes, and which are usually controlled by only a few genes; various coat colors and the horned trait in domestic animals are good examples.

Quantitative traits: Traits that do not have a sharp distinction between phenotypes, and usually require some kind of measuring tool to make the distinctions; these traits are normally controlled by many pairs of genes; examples are growth rate, milk production, and carcass quality.

Quarantine: The segregation of a diseased or exposed animal from other animals in order to prevent the spread of a contagious disease.

Rabies: Hydrophobia, a viral disease of the brain and nervous system; it results in an inability to swallow.

Rack: A rapid, four-beat gait of a horse that requires a considerable amount of training to perform; performed by American Saddle Horses.

Ram: A male sheep.

Random mating: A mating system with no selection where every male has an equal chance of mating with every female.

Ration: The amount of feed consumed or provided within a 24-hour period.

Recessive: Refers to a gene whose expression can be modified or covered up by an allele.

Recombination: The formation of genotypes resulting from the various possible unions between the gametes of the parents.

Rectum: The terminal part of the large intestine.

Red meat: Refers generally to the meat of cattle, sheep, hogs, and goats as opposed to that of poultry.

Reflex: An involuntary response to a stimulus.

Regurgitation: A process whereby the contents of the stomach are carried through the esophagus to the mouth.

Relationship: A genetic term that refers to the proportion of genes that any two individuals have which are the same.

Relaxin: An ovarian hormone produced at the time of parturition that is thought to aid in the relaxation of the birth canal.

Retained placenta: Fetal membranes that are not expelled following parturition.

Rete testis: A network of tubules located inside the testis in the mediastinum connecting the seminiferous tubules to the efferent ducts.

Reticulum: One of the smaller of the four compartments of the ruminant stomach located next to the rumen.

Rickettsea: A group of microorganisms that are intermediate in size to bacteria and viruses; they survive by reproducing in the cells of larger organisms in much the same way as do viruses.

Ridgling: Another term for a cryptorchid.

Rigor mortis: The stiffness that occurs in the skeletal muscles of an animal shortly after death.

Roan: A color pattern in cattle and horses produced by the interspersion of white hairs among hairs of a contrasting color.

Rooster: A mature male chicken; another term for cock.

Roughage: A kind of feedstuff, consisting of the leaves and stems of plants, that is relatively high in fiber content; examples are hay, pasture, and silage.

Rumen: The largest compartment of the stomach of cattle, sheep, and goats and their relatives; a large amount of bacterial fermentation of feed materials occurs in the rumen; also called the paunch.

Ruminant: An animal with a rumen and other stomach compartments.

Rumination: A process in animals with a rumen whereby the food is regurgitated as a bolus (cud), rechewed, and reswallowed.

Sagittal: Refers to a plane or section through the body that is parallel to the medial plane.

Saturated fat: A fat whose carbon atoms are associated with the maximum number of hydrogen atoms; no double bonds exist.

Scoured wool: Wool that has been treated and cleaned to remove the grease and other substances.

Scours: An abnormal diarrhea in animals.

Scurs: A hornlike substance that grows from the skin at the location of the horn pits on polled animals.

Sebaceous gland: A skin gland that secretes a fattylike substance.

Second meiotic division: The second of two divisions in the process of producing sex cells; the two chromatids of each chromosome separate to produce a chromosome for each of the two daughter cells of the second division, in this regard it is similar to the single division of mitosis.

Segregation of genes: The separation of genes on two homologous chromosomes and their distribution to separate gametes during gametogenesis; genes that once existed in pairs in a cell will become separated and distributed to different gametes.

Selection: A process that causes or allows certain animals to mate and produce progeny, and prevents other animals from doing so.

Selection differential: The difference between the average of a selected population and the average of the population from which they came.

Semen: A fluid substance produced by the male reproductive system containing spermatozoa suspended in secretions of the accessory glands.

Sensory: Referring to sensations, the ability to convert stimuli to nerve impulses.

Septum: A separating wall or membrane.

Serosa: The outer covering of the internal organs; it secretes a salivalike lubricating fluid that prevents adhesion of organs one to another.

Serum: The clear or yellowish plasmalike fluid that results after blood coagulates.

Sex chromosomes: A pair of chromosomes in animals that determines the sex of the progeny depending upon which one is distributed; one sex usually has two of the same kind of sex chromosome in its cells while the other sex has two kinds of sex chromosomes; in mammals the female is XX and the male is XY; in birds, the male is ZZ (or XX) and the female is ZW (or XY).

Sex-influenced traits: Traits affected by autosomal genes showing dominance and recessiveness between the alleles; however, the allele that is dominant in one sex will be recessive in the other.

Sex-limited traits: Traits that are expressed only in one sex although the genes for the trait are carried by both sexes.

Sex-linked genes: Genes that are found on the sex chromosomes.

Shank: That part of the leg of a chicken between the hock and the foot; a cut of meat from the lower part of the foreleg or hindleg of a beef or lamb carcass.

Shoat: A young pig of either sex.

Sib: Brother or sister.

Sigmoid flexure: An S-shaped fold in the penis of a bull, ram or boar that straightens during erection and allows it to extend from the sheath.

Silage: Roughage that is cut green containing approximately 65 to 70 percent moisture and stored in an airtight structure called a silo.

Sinus: A cavity inside a bone.

Sire: The male parent; the father.

Skewbald: A color pattern in horses involving white spots on any color but black.

Snood: The relatively long fleshy extension at the base of the beak of a turkey.

Somatic: Refers to body tissues or cells in contrast to sex cells.

Sorrel: A lighter shade of chestnut in horses.

Sow: A mature female hog.

Spay: To surgically remove the ovaries.

Species: A group of animals with certain common characteristics that when mated together will produce fertile offspring.

Sperm: A spermatozoon; the male sex cell.

Spermatid: A haploid cell produced from the second division of meiosis in spermatogenesis that has not yet undergone the changes to form a sperm cell.

Spermatogenesis: The process in males of producing spermatozoa.

Spermatogonium: A primary germ cell in the testis that will undergo spermatogenesis to produce spermatozoa.

Spermiogenesis: That part of the process of spermatogenesis involving the changes that permits spermatids to become spermatozoa.

Sphincter: A circular or ringlike muscle in a tubular organ that acts as a valve to close the opening in the tube.

Spindle: The array of what appears to be fibers extending from one centriole to the other in a cell in the prophase or metaphase of division.

Spirochete: A spiral-shaped bacterium.

Spore: A reproductive cell of some microorganisms.

Springer: A cow in advanced stages of pregnancy.

Spur: A sharp toelike projection on the back of the shank of male birds.

Stag: A male animal that has been castrated after reaching sexual maturity.

Stallion: A mature male horse.

Staphylococcus: Spherical or dot-shaped bacteria that occur in clusters.

Starch: A kind of carbohydrate manufactured by plants and stored in the seeds, roots, and fruit as a reserve energy supply; the major component of livestock feeds.

Steer: A male bovine animal castrated before reaching puberty.

Stifle: The joint in the hind leg of a quadruped formed by the femur, tibia, patella, and sometimes the fibula.

Stilbestrol: Diethylstilbestrol, a synthetic estrogenic hormone.

Stillborn: Born dead.

Stover: The mature stalks and leaves of plants, such as corn or sorghum, after the grain has been removed.

Strain: A group of animals within a breed having certain characteristics that distinguish it from others of the breed; family.

Strain cross: A cross between members of two different strains.

Streptococcus: Spherical or dot-shaped bacteria that occur in chainlike colonies.

Stress: Abnormal or adverse conditions and factors to which an animal cannot adapt or adjust to satisfactorily, resulting in physiological tension and possible disease; the factors may be physical, chemical, or psychological.

Stud: A group of male animals kept for breeding purposes, or the place or establishment where the breeding takes place, as in the case of a bull stud or dog stud.

Sugar: A sweet, water-soluble carbohydrate of relatively low molecular weight.

Suint: Solid deposits from the perspiration of sheep found in the wool.

Superfetation: The presence of two fetuses in the uterus that resulted from fertilizations during two different estrous cycles.

Superovulation: The stimulation of more than the usual number of ovulations during a single estrous cycle due to the injection of certain hormones.

Supplement: A feed or feed additive given to animals to provide nutrients such as protein, minerals, or vitamins that were deficient in the basic ration.

Switch: The brush of hair at the end of a cow's tail.

Symbiosis: The situation where two organisms that are not alike live together in such a way that the association is beneficial to both.

Synapsis: The coming together or pairing of homologous chromosomes during the prophase of the first division of meiosis.

Syndrome: A group of symptoms occurring together that characterize some abnormality or disease.

Tack: The riding equipment of a horse, such as the saddle and bridle.

Tallow: The fat of cattle and sheep.

Tankage: A relatively high protein feed containing meat and bone by-products.

Taxonomy: The science that deals with the orderly classification of living organisms.

TDN: *See* Total digestible nutrients.

Tease: The stimulation of an animal to accept copulation.

Telophase: The phase of cell division between anaphase and the complete separation of the two daughter cells; includes the formation of the nuclear membrane and the return of the chromosomes to long, threadlike and indistinguishable structure.

Tendon: A kind of connective tissue that attaches muscle to bone.

Testcross: A kind of genetic cross involving one individual expressing a dominant trait and one expressing the recessive trait, the purpose of the cross being to determine whether the individual expressing the dominant trait is heterozygous or homozygous.

Testis: The primary sex organ of the male, the source of the male gametes and the male sex hormone.

Testosterone: A hormone produced by the interstitial cells of the testes that functions in stimulating male sex drive, masculine characteristics, development of the male reproductive tract, and spermatogenesis.

Tetanus: A disease of the nervous system caused by an anaerobic bacillus that results in an inability to control certain muscles, particularly those in the region of the neck and jaw; also called lockjaw.

Tetany: Sustained or spasmodic muscle contractions.

Tetrad: A pair of synapsed homologous chromosomes observed during the prophase or metaphase of the first division of meiosis.

Theca: The membranes that form the wall of the ovarian follicle.

Therm: A megacalorie, equivalent to 1000 kcal or 1,000,000 cal.

Thoracic: Refers to the chest, the chest cavity, or to the vertebral column in the region of the chest.

Thyroid: An endocrine gland located in the neck that secretes thyroxine.

Thyroid-stimulating hormone (TSH): A hormone produced by the anterior pituitary gland that stimulates the thyroid to produce thyroxine.

Thyroxine: A hormone of the thyroid gland that controls the metabolic activity of cells.

Tissue: Groups of cells working together to carry out a common function, such as muscle tissue, connective tissue, and epithelial tissue.

Tom: A male turkey or cat.

Total digestible nutrients (TDN): A standard evaluation of a feed that shows its usefulness as a source of digestible carbohydrates, fats, and protein, or any combination of the three.

Toxin: A poisonous or toxic product of certain microorganisms.

Trachea: The windpipe, extending from the throat to the bronchi.

Trait: Any observable feature or characteristic of an animal.

Transverse plane: A section or plane running perpendicular to the axis of the body or organ.

Trihybrid: An individual that is heterozygous for three pairs of genes.

Triploidy: A condition where the cells possess three sets of homologous chromosomes rather than two; 3n.

Trot: A two-beat gait in horses where the diagonal legs move and strike the ground together.

TSH: *See* Thyroid-stimulating hormone.

Tuberculosis: A disease caused by a bacillus affecting the lungs and certain other tissues.

Tunica albugenia: A capsulelike covering of an organ, in particular, the testes.

Tunica dartos: The layer of muscle in the scrotum that helps control the distance between the testes and the body which in turn regulates the temperature of the testes at a point several degrees below body temperature.

Tunica vaginalis: The lining of the scrotum nearest the testes; it is an extension of the lining of the abdominal cavity that passes through the inguinal canal and into the scrotum.

Udder: The mammary glands, including the teats or nipples of farm mammals.

Umbilical cord: The part of the fetal membranes that connects to the navel of the fetus; it carries the blood vessels that transports fetal blood to and from the placenta.

Undulant fever: A disease in humans caused by the bacteria that cause brucellosis in domestic animals.

Urea: A nonprotein, organic substance containing approximately 45 percent nitrogen which is sometimes used as a feed additive in ruminants.

Uterus: The part of the reproductive tract of female mammals in which the fetus develops during pregnancy.

Vaccination: A process of injecting controlled amounts of microorganisms or microorganism products into animals in an effort to prevent that animal from contracting a disease caused by that particular organism; the substance injected is called an antigen; it stimulates the production of antibodies that provides some protection to the host from the invading organisms.

Vaccine: A substance that contains live, modified, or dead organisms or their products which is injected into an animal in an attempt to protect the host from a disease caused by that particular organism.

Variance: A statistical measure of the amount of variation that is observed within or among a group of animals.

Variation: A measure of the differences observed among animals.

Vascular: Referring to blood vessels.

Vasectomy: A surgical method of sterilization in males whereby the vasa deferentia are cut to prevent the spermatozoa from being transported from the testes during ejaculation; the ejaculate contains only the seminal plasma.

Vasotocin: An oxytocinlike hormone produced in birds that stimulates oviposition.

Veal: The meat from calves slaughtered before they are three months old.

Vector: An organism such as an insect or arachnid, that carries a disease-producing microorganism.

Vena cava :   The large veins that carry blood into the right atrium of the heart.

Ventral: Toward the belly or lower surface of the body; opposite of dorsal.

Ventricle: One of the two pumping chambers of the heart.

VFA: *See* Volatile fatty acids.

Vibriosis: A bacterial venereal disease in cattle that can cause abortions.

Villi: Microscopic, hairlike extensions or projections of the inner lining of the digestive tract or of the placenta.

Virulent: Refers to an organism that is harmful and can cause a disease with relative ease.

Virus: One of a group of very small microscopic living agents that do not have all of the parts of a typical cell, and must reproduce inside of another living cell.

Viscera: A collective term that refers to all the internal organs in the thoracic and abdominal cavities.

Vitamins: A variable group of organic substances that function as catalysts, or as components of catalysts, and which are usually required in relatively small amounts.

Vitelline membrane: The thin membrane located inside the zona pellucida of an ovum that contains the cytoplasm or vitellus.

Vitellus: The cytoplasm of a mammalian ovum; it contains relatively large amounts of nutritive material called yolk or deutoplasm.

Viviparous: The bringing forth of living offspring from the body as in mammals as opposed to the laying and hatching of eggs.

Volatile fatty acids (VFA):   A group of low molecular weight acids that form gases rather easily, and are produced by microbial action in the rumen; examples are acetic, propionic, and butyric acid.

Vomit: To forcibly expel stomach contents through the esophagus and into the mouth.

Vulva: The fleshy folds of tissue that cover the terminal or external end of the vagina.

Wattle: One of the two, thin, leaflike structures suspended from the upper part of the neck of a chicken; they are made of the same kind of tissue as the comb.

Weanling: An animal that has been recently weaned.

Wether: A male sheep or goat that has been castrated before reaching puberty.

Whelp: To give birth to a litter of pups; may also refer to the pup.

Whey: The watery portion of milk that remains after the curd and cream have been removed; it contains some protein, sugar, and other soluble materials.

Withers: The ridge at the base of the neck and the point of the shoulders in farm animals.

Worsted wool: A smooth compact yarn made from long wool fibers and used for smooth fabrics such as gabardine.

Yean: To give birth to, particularly in sheep and goats.

Yeld mare: A mare that is not lactating; a dry mare.

Yield: In meat animals, referring to the amount of lean meat produced in a carcass.

Yolk: The nutritive material in an egg or ovum that provides sustenance to the developing embryo; the yellow part of the egg of a chicken.

Zebu: The kind of cattle originating from India from which the Brahman breed was developed; common characteristics include the shoulder hump, the drooping ears, and the loose pendulous dewlap.

Zein: The protein in the corn grain.

Zona pellucida: The relatively thick covering or membrane that forms the outer surface of a mammalian ovum.

Zygote: The diploid cell formed from the union of the sperm cell with an ovum.

Bibliography

GENERAL

Acker, Duane. 1971. *Animal Science and Industry*, 2nd ed. Englewood Cliffs, N.J.: Prentice-Hall, Inc.

Becker, Raymond B. 1973. *Dairy Cattle Breeds*. Gainesville: University of Florida Press.

Blakely, James and David H. Bade. 1981. *The Science of Animal Husbandry*, 3rd ed. Reston, Va: Reston Publishing Co.

Bogart, Ralph. 1977. *Scientific Farm Animal Production*. Minneapolis: Burgess Publishing Co.

Briggs, H. M. and D. M. Briggs. 1980. *Modern Breeds of Livestock*, 4th ed. Toronto: The Macmillan Company.

Campbell, John R. and John F. Lasley. 1975. *The Science of Animals That Serve Mankind*, 2nd ed. New York: McGraw-Hill Book Co.

Campbell, John R. and R. T. Marshall. 1975. *The Science of Producing Milk for Man*. New York: McGraw-Hill Book Co.

Cole, H. H. and W. N. Garrett. 1980. *Animal Agriculture*, 2nd ed. San Francisco: W. H. Freeman and Company.

454

Diggins, R. V. and V. W. Christensen. 1975. *Livestock and Poultry Production*. Englewood Cliffs, N.J.: Prentice-Hall, Inc.

Ensminger, M. E. 1977. *Animal Science*, 7th ed. Danville, Ill.: Interstate Printers and Publishers, Inc.

Evans, J. W., A. Borton, H. F. Hintz, and L. D. Van Vleck. 1977. *The Horse*. San Francisco: W. H. Freeman and Company.

May, Cheryl. 1981. *Cattle Management*. Reston, Va: Reston Publishing Co.

ANATOMY AND PHYSIOLOGY

Bone, Jesse F. 1979. *Animal Anatomy and Physiology*. Reston, Va.: Reston Publishing Co.

Breazile, J. E. 1971. *Textbook of Veterinary Physiology*. Philadelphia: Lea and Febiger.

Frandson, R. D. 1981. *Anatomy and Physiology of Farm Animals*, 3rd ed. Philadelphia: Lea and Febiger.

Swenson, Melvin J. 1977. *Dukes' Physiology of Domestic Animals*, 9th ed. Ithaca, N.Y.: Comstock Publishing Associates.

ANIMAL HEALTH

Berrier, Harry H. 1977. *Animal Sanitation and Disease Prevention*, 2nd ed. Dubuque, Iowa: Kendall/Hunt Publishing Co.

Blakely, James and D. D. King. 1978. *The Brass Tacks of Animal Health*. St. Louis: Doane Agricultural Services, Inc.

Bruner, D. W. and J. H. Gillespie. 1973. *Hagan's Infectious Diseases of Domestic Animals*, 6th ed. Ithaca, N.Y.: Cornell University Press.

Dykstra, R. R. 1961. *Animal Sanitation and Disease Control*, 6th ed. Danville, Ill.: Interstate Printers and Publishers, Inc.

Hungerford, T. G. 1967. *Diseases of Livestock*, 6th ed. Sydney, Australia: Angus and Robertson, Ltd.

USDA 1956. *Animal Diseases, Yearbook of Agriculture*. Washington, D. C.: U.S. Department of Agriculture.

PHYSIOLOGY OF REPRODUCTION

Fuquay, John W. and Joe H. Bearden. 1980. *Applied Animal Reproduction*. Reston, Va.: Reston Publishing Co.

Hafez, E. S. E. 1970. *Reproduction and Breeding Techniques for Laboratory Animals*. Philadelphia: Lea and Febiger.

——. 1980. *Reproduction in Farm Animals*, 4th ed. Philadelphia: Lea and Febiger.

McDonald, L. E. 1975. *Veterinary Endocrinology and Reproduction*, 2nd ed. Philadelphia: Lea and Febiger.

Nalbandov, A. V. 1976. *Reproductive Physiology of Mammals and Birds*, 3rd ed. San Francisco: W. H. Freeman and Company.

Salisbury, G. W., N. L. VanDenmark, and J. R. Lodge. 1978. *Physiology of Reproduction and Artificial Insemination of Cattle*, 2nd ed. San Francisco: W. H. Freeman and Company.

GENETICS AND ANIMAL BREEDING

Bogart, Ralph. 1959. *Improvement of Livestock*. New York: The Macmillan Company.

Burns, G. W. 1972. *The Science of Genetics*. New York: The Macmillan Company.

Hutt, F. B. 1982. *Animal Genetics*, 2nd ed. New York: John Wiley & Sons, Inc.

Johansson, I. and J. Rendel. 1968. *Genetics and Animal Breeding*. San Francisco: W. H. Freeman and Company.

Jones, W. E. and Ralph Bogart. 1971. *Genetics of the Horse*, 2nd ed. Fort Collins, Colo.: Caballus Publishers.

King, Robert C. 1975. *Handbook of Genetics Volume 4: Vertebrates of Genetic Interest*. New York: Plenum Press.

Lasley, J. F. 1978. *Genetics of Livestock Improvement*, 3rd ed. Englewood Cliffs, N.J.: Prentice-Hall, Inc.

Lush, J. L. 1963. *Animal Breeding Plans*. Ames, Iowa: Collegiate Press, Inc.

Snyder, Laurence H. and Paul R. David. 1957. *The Principles of Heredity*, 5th ed. Boston: D. C. Heath and Co.

Warwick, E. J. and Legates, J. E. 1979. *Breeding and Improvement of Farm Animals*, 7th ed. New York: McGraw-Hill Book Co.

ANIMAL NUTRITION

Church, D. C. and W. G. Pond. 1974. *Basic Animal Nutrition and Feeding*. Corvalis, Ore.: O and B Books.

Crampton, E. W. and L. E. Harris. 1969. *Applied Animal Nutrition*, 2nd ed. San Francisco: W. H. Freeman and Company.

Cullison, A. 1982. *Feeds and Feeding*, 3rd ed. Reston, Va.: Reston Publishing Co.

Ensminger, M. E. and C. G. Olentine. 1978. *Feeds and Nutrition*. Clovis, Calif. Ensminger Publishing Co.

Jurgens, M. H. 1978. *Animal Feeds and Nutrition*, 4th ed. Dubuque, Iowa: Kendall/Hunt Publishing Co.

Maynard, L. A. and J. K. Loosli. 1069. *Animal Nutrition*, 6th ed. New York: McGraw-Hill Book Co.

Morrison, F. B. 1956. *Feeds and Feeding*, 22nd ed. Ithaca, N.Y.: The Morrison Publishing Co.

ANIMAL PRODUCTS

AMI Center for Continuing Education, American Meat Institute. 1960. *The Science of Meat and Meat Products*. San Francisco: W. H. Freeman and Company.

Aykroyd, W. R. 1964. *Food for Man*. London: Pergamon Press.

Campbell, John H. and R. T. Marshall. 1975. *The Science of Providing Milk for Man*. San Francisco: McGraw-Hill Book Co.

Ensminger, M. E. 1970. *Sheep and Wool Science*, 4th ed. Danville, Ill.: Interstate Printers and Publishers, Inc.

Forrest, John C., Elton D. Aberle, Harold B. Hedrick, Max D. Judge, and Robert A. Merkle. 1975. *Principles of Meat Science.* San Francisco: W. H. Freeman and Company.

Henrickson, R. L. 1978. *Meat, Poultry, and Seafood Technology.* Englewood Cliffs, N.J.: Prentice-Hall, Inc.

Lampert, L. N. 1975. *Modern Dairy Products*, 3rd ed. New York: Chemical Publishing Co.

McCoy, J. H. 1972. *Livestock and Meat Marketing.* Westport, Conn.: AVI Publishing Company.

McFadden, W. D. 1967. *Wool Science.* Boulder, Colo.: Pruett Press, Inc.

Nesheim, M. C., R. E. Austic, and L. E. Card. 1979. *Poultry Production*, 12 th ed. Philadelphia: Lea and Febiger.

Romans, J. R. and P. T. Ziegler. 1974. *The Meat We Eat*, 10th ed. Danville, Ill.: Interstate Printers and Publishers, Inc.

Schmidt, G. H. and L. D. Van Vleck. 1974. *Principles of Dairy Science.* San Francisco: W. H. Freeman and Company.

USDA. 1959. *Food: The Yearbook of Agriculture.* Washington, D. C.: U.S. Department of Agriculture.

Index